Instructor's Manual to Accompany Linear Algebra and Ordinary Differential Equations

ALAN JEFFREY

Blackwell Scientific Publications

Instructor's Manual to Accompany
Linear Algebra and
Ordinary Differential Equations

ALAN JEFFREY
Professor of Engineering Mathematics,
University of Newcastle upon Tyne, and
Adjunct Professor of Mathematics,
University of Delaware

CRC Press
Taylor & Francis Group
Boca Raton London New York

CRC Press is an imprint of the
Taylor & Francis Group, an informa business

CRC Press

way NW, Suite 300

'42

ress

ntific Publications, Inc.
of Taylor & Francis Group, an Informa business

. Government works

54-6 (hbk)
64-6 (ebk)

is Web site at http://www.taylorandfrancis.com and the
tp://www.crcpress.com

CONTENTS

FOREWORD

This manual, which has been prepared for the benefit of instructors, gives answers to the even numbered problems in Linear Algebra and Ordinary Differential Equations, comprising Volume 1 of Advanced Engineering Mathematics.

In the case of straightforward problems only the answer itself has been given, though where the problems are more difficult outline solutions have been provided in which each key step is described in detail. For convenience of reference, the page numbers in the manual on which the answers to each Section are to be found have been listed in the Contents at the front against the name of the corresponding Section in the book. Not every Section of the book has an associated problem set.

It is a pleasure to record my gratitude to my secretary Mrs. Lynn Kelly for preparing the typescript from which this manual has been prepared.

Alan Jeffrey

November 1989

Chapter 1. Review of Topics from Analysis.

Section 1.2

4. $x \le -1, \ x \ge 7$

6. $x < -4, \ x > 2$

8. $1 < x < 3$ and $3 < x < 5$

10. $-2 < x \le -\frac{5}{4}$ and $\frac{7}{4} \le x < \frac{5}{2}$

12. $(a - b)^2 \ge 0$, so

```
—————●                    ●—————
     -1                    7

—————○          ○—————
     -4          2

—————○     ○     ○—————
     1     3     5

——○——————●———————●——————○—
 -2    -5/4    7/4    5/2
```

$a^2 - 2ab + b^2 \ge 0$ or $a^2 + b^2 \ge 2ab$.   Result follows by setting

$a = \sqrt{(x_1/2)}$, $b = \sqrt{(x_2/2)}$.

14. $\left[1 + \frac{1}{n}\right]^n = 1 + n\left[\frac{1}{n}\right] + \frac{n(n-1)}{2!} \left[\frac{1}{n}\right]^2 + \ldots + \left[\frac{1}{n}\right]^n$

$< 1 + 1 + \frac{n^2}{2!} \left[\frac{1}{n}\right]^2 + \frac{n^3}{3!} \left[\frac{1}{n}\right]^3 + \ldots + \frac{1}{n!} = 1 + \sum_{r=1}^{n} \frac{1}{r!}$.

All terms are positive so

$2 = 1 + n\left[\frac{1}{n}\right] < \left[1 + \frac{1}{n}\right]^n$.   Hence the result.

16. Result follows by setting $b_1 = b_2 = \ldots = b_n = 1$.

18. If $0 < x < 1$ then $(x - 1) < 0$ and as $x^{n-1} + x^{n-2} + \ldots + x + 1 < n$ the second factor is also negative.   Hence RHS > 0 for $0 < x < 1$, when $x^n - nx + n - 1 \ge 0$ follows at once.   Similarly, if $x > 1$ both factors are positive and the result again follows.   Result is trivial for $x = 0$ and $x = 1$.

Section 1.3

10. True

12. True

14. True

20. $a_n = \frac{(-1)^n a_0}{(2n + 1)!}$

22. $a_n = \frac{(-1)^n a_0}{2^{\frac{1}{2}n(n+1)} n!}$

26. Argue as in Problem 25.   (i) 7/16 ($r = 6$) and (ii) $-21/2$ ($r = 5$).

1

Section 1.4

2. $1 + 3x + \dfrac{2}{x - 1} + \dfrac{1}{x + 2}$

4. $\dfrac{2}{(x^2 + 1)^2} + \dfrac{x}{x^2 + x + 2} \equiv \dfrac{2}{(x^2 + 1)^2} + \dfrac{x}{(x + \frac{1}{2})^2 + \frac{7}{4}}$

6. $\dfrac{2}{(x - 1)^2} + \dfrac{3}{x + 2}$

8. $\dfrac{2}{x} + \dfrac{3}{x + 3} + \dfrac{x}{(x - 1)^2 + 3}$

10. $\dfrac{1}{4x} + \dfrac{1}{2x^2} + \dfrac{1}{x + 1} - \dfrac{1}{4}\dfrac{1}{x + 2}$

Section 1.6

2. $0, \sqrt{2} + i\sqrt{2}, (\sqrt{2} - i\sqrt{2})/2, i, i$

4. $-7 + 24i$ 6. $3/25$

8. -3 10. $1/2$

22. Center 3, radius 1, center $-2 + i$, radius 2, center $2i$, radius 3

24. $2i, -2i; \quad (z - 2i)(z + 2i)$

26. $-3 + i\sqrt{3}, -3 - i\sqrt{3}; \quad (z + 3 - i\sqrt{3})(z + 3 + i\sqrt{3})$

28. $2, -2, 2i; \quad (z - 2)(z + 2)(z - 2i)(z + 2i)$

30. $1, -2 + i\sqrt{2}, -2 - i\sqrt{2}$

32. $-1 + 2i, -1 - 2i, 4i, -4i$

34. $-i, 1, -2i, 2i$. As the coefficients are complex, Theorem 1.4 does not apply.

Section 1.7

2. $2, 3\pi/4, 3\pi/4 + 2k\pi$ 4. $2, \pi/2, \pi/2 + 2k\pi$

6. $\left[\cos \dfrac{\pi}{4} + i \sin \dfrac{\pi}{4}\right]$ 8. $3\left[\cos \dfrac{\pi}{2} + i \sin \dfrac{\pi}{2}\right]$

10. $2, \left[\cos \dfrac{\pi}{2} + i \sin \dfrac{\pi}{2}\right]$ 12. $\pi/2$

14. $\sin 4\theta = 4 \cos^3 \theta \sin \theta - 4 \cos \theta \sin^3\theta = \cos \theta(4 \sin \theta - 8 \sin^3\theta)$

     $\cos 4\theta = \cos^4\theta - 6 \cos^2\theta \sin^2\theta + \sin^4\theta = 8 \cos^4\theta - 8 \cos^2\theta + 1$

2

16. $2^{25/2} \left[\cos \frac{3\pi}{4} + i \sin \frac{3\pi}{4} \right]$

18. $\bar{z} = 5(\cos \alpha - i \sin \alpha)$, $1/z = \frac{1}{5}(\cos \alpha - i \sin \alpha)$, $(\bar{z})^2 = 25(\cos 2\alpha - i \sin 2\alpha)$, $1/z^3 = \frac{1}{125}$ $(\cos 3\alpha - i \sin 3\alpha)$ with $\alpha = $ arc tan (4/3) (acute angle).

22. $2^{7/6} \exp \left[\left[\frac{8k + 1}{12} \right] \pi i \right]$ $(k = 0, 1, 2)$.

26. $16^{2/5} \exp \left[\left[\frac{5 + 6k}{15} \right] \pi i \right]$, $(k = 0, 1, 2, 3, 4)$.

30. $3^{1/2} e^{\frac{-\pi i}{4}}$ (twice) , $3^{1/2} e^{\frac{3\pi i}{4}}$ (twice) .

34. Proceed as indicated in the problem.

Section 1.8

2. $F(x) = (x - 1)e^x - e^2$

4. $F(x) = (a^x - a^3)/\ln a$

6. $F(x) = x - \pi/4$, $(\pi/4 \leq x \leq \pi/2)$

8. $F(x) = \frac{1}{x} [\cos(x^2) - \cos(x + x^3)]$, $(x > 0)$

10. $F'(x) = - (1 + x^6)^{1/2}$

12. $F'(x) = \ln x + \frac{1}{x} + 1$, $(x > 0)$

14. $F'(x) = \frac{1}{2} (1 + 4x) \sinh (x + 2x^2) - \frac{3}{2} \sinh 3x$

18. $\xi = 0$ is unique.

20. ξ is not unique, for $\xi = \pm \frac{1}{2}$.

22. (a) $\xi = 2(a^2 + ab + b^2)/[3(a + b)]$ is unique.

(b) $\xi = $ arc cos $\frac{2}{\pi}$, $0 < \xi < \frac{\pi}{2}$ is unique.

24. $I \leq \frac{\pi}{4} \sin 1 \simeq 0.21\pi$

26. π

28. $\ln (2 + \sqrt{3})$

30. divergent

32. divergent

34. divergent

36. divergent

38. convergent; compare with $\int_{-\pi/2}^{0} dx/x^{2/3}$

40. $\pi/2$

42. π

44. divergent

46. $\pi/\sqrt{5}$

48. $\frac{1}{3} + \frac{1}{4} \ln 3$

50. convergent

52. divergent

54. convergent if $\mu > 1$, divergent $\mu \leq 1$.

56. $dx/dt = \omega(a^2 - x^2)^{\frac{1}{2}}$, where the positive square root is taken because time must increase. Thus

$$\int_0^{T/2} dt = \frac{1}{\omega} \int_{-a}^a \frac{dx}{(a^2 - x^2)^{\frac{1}{2}}} \quad \text{from which it follows that } T = 2\pi/\omega.$$

58. To prove (ii), for example, use

$$\int_a^b f(x)\ g(x)\ dx = \lim_{n \to \infty} \sum_{i=1}^n f(x_i)\ g(x_i)\ \Delta_i,$$

so

$$m \lim_{n \to \infty} \sum_{i=1}^n g(x_i)\ \Delta_i \leq \int_a^b f(x)\ g(x)\ dx \leq M \lim_{n \to \infty} \sum_{i=1}^n g(xi)\ \Delta_i.$$

60. Express $\Gamma(x)$ as the sum of two improper integrals

$$\Gamma(x) = \int_0^1 t^{x-1} e^{-t} dt + \int_1^\infty t^{x-1} e^{-t} dt.$$

Compare the first integral with $\int_0^1 (1/t^{x-1})\ dt$. Apply Theorem 1.8 to the second integral with the upper limit replaced by $b > 1$, allow $b \to +\infty$ and use the fact that $\lim_{t \to \infty} t^\alpha e^{-t} = 0$ for any α.

The convexity follows by differentiating under the integral sign twice with respect to x to show

$$\Gamma''(x) = \int_0^\infty e^{-t} t^{x-1} (\ln t)^2\ dt > 0.$$

62. $x^\mu f(x) \geq A > 0$, so $f(x) \geq A/x^\mu$.

Now compare $\int_a^\infty f(x)\ dx$ with $\int_a^\infty \frac{A}{x^\mu}\ dx.$

4

64. (i) $\displaystyle\int_{\pi/2}^{\infty} \frac{\sin x}{x}\, dx = -\int_{\pi/2}^{\infty} \frac{\cos x}{x^2}\, dx$, which is convergent since

$(\cos x)/x^2 \leq 1/x^2$ and $\displaystyle\int_{\pi/2}^{\infty} \frac{dx}{x^2}$ converges.

(ii) $\displaystyle I_3 = \lim_{R \to \infty} \int_{\pi/2}^{R} \frac{1 - \cos 2x}{2x}\, dx = \frac{1}{2}\lim_{R \to \infty} \ln R - \frac{1}{2} \ln \frac{\pi}{2}$

$\displaystyle - \frac{1}{2}\int_{\pi/2}^{\infty} \frac{\cos 2x}{x}\, dx.$

The limit is infinite, and the integral is seen to be convergent by comparison

with $\displaystyle\int_{a}^{\infty} \frac{\cos x}{x}\, dx$, which is convergent.

66. (i) For example, let $f(x) = (\sin x)/x$ and

$$g(x) = \begin{cases} 1 & , \quad 2\pi \leq x \leq (2 + 1)\pi \\ -1 & , \quad (2 + 1)\pi < x < (2 + 2)\pi. \end{cases}$$

Then

$$\int_{0}^{\infty} f(x)\, dx = \int_{0}^{\infty} \frac{\sin x}{x}\, dx$$

is the Dirichlet integral which is convergent, but

$$\int_{0}^{\infty} f(x)\, g(x)\, dx = \int_{0}^{\infty} \left|\frac{\sin x}{x}\right|\, dx,$$

is divergent (Prob. 64).

(ii) If $g(x)$ is bounded then $|g(x)| < C$, and so $|f(x)\, g(x)| < C|f(x)|$.

The result then follows from the fact that $\displaystyle\int_{a}^{\infty} f(x)\, dx$ is absolutely convergent.

Section 1.9

2. $\lambda_1 = 1, \lambda_2 = 2$; $u_n = A + B.2^n$; $u_n = 2^{n+1} - 1$ for $n = 0, 1, 2, \ldots$.

4. $\lambda_1 = \lambda_2 = 2$; $u_n = A.2^n + Bn2^{n-1}$; $u_n = 2^{n-1}(2 - n)$ for $n = 0, 1, 2, \ldots$.

6. $\lambda_1 = 2^{1/2}e^{\pi i/4}$, $\lambda_2 = 2^{1/2}e^{-\pi i/4}$; $u_n = 2^{n/2}\left[A \cos \frac{n\pi}{4} + B \sin \frac{n\pi}{4}\right]$;

   $u_n = 2^{(n-1)/2} \sin \frac{n\pi}{4}$ for $n = 0, 1, 2, \dots$.

8. $\lambda_1 = 3$, $\lambda_2 = -3$; $u_n = A.3^n + B(-3)^n - 1$;

   $u_n = 3^n + (-3)^n - 1$ for $n = 0, 1, 2, \dots$.

10. $\lambda_1 = 2e^{i\pi/2}$, $\lambda_2 = 2e^{-i\pi/2}$; $u_n = 2^n \left[A \cos \frac{n\pi}{2} + B \sin \frac{n\pi}{2}\right]$ +

   $\frac{2}{5} + \left[\frac{1}{13}\right]3^n$;

   initial conditions are satisfied if in u_n above $A = -31/65$, $B = 12/65$.

Chapter 2.   Algebra of Vectors.

Section 2.3

2.   Self–checking

4.   The sum of the lengths of any two sides of a triangle exceeds the length of the third side.   Equality only when **a** and **b** are parallel and have same sense.

6.   **a** + **b** and **a** − **b** are parallel to the diagonals of a rhombus, and these are mutually perpendicular.

8.   In general **a** + **b** + **c** will form an open 3 sided polygon.   If **a** + **b** + **c** = **0** this polygon closes and becomes a triangle which must then lie in a plane.

10.   Show that the position vector of the point R on each median is identical.

12.   Use the result $v_w = v_b - v_{bw}$, where v_b, v_w are the respective velocities of the boat and the water relative to the earth and v_{bw} is the velocity of the boat relative to the water. Water speed $5\sqrt{2}$ m/sec; direction due east.

14.   Forces are vectors, so the resultant force is their vector sum, and it will be directed from the initial point of F_1 to the terminal point of F_n.   The line of action of R will be through P because all the forces act through P.   (a) R = 0 if polygon is closed   (b) if R ≠ 0, equilibrium produced by addition of force −R with its line of action through P.

Section 2.4

2.   (a) Q(3, 1, 2)   (b) Q(−2, 4, −3)   (c) P(1, 0, 5)   (d) P(2, 5, 8)

4. $4i + 8j + 12k$, $i + \frac{1}{3}j + \frac{2}{3}k$, $4i − 12j + 2k$

6. $−i + 4j + 3k$, $3i + 3k$

8. $3\sqrt{2}$, $\sqrt{14} + \sqrt{2}$, $\sqrt{2} − \sqrt{14}$, $\sqrt{14} − \sqrt{2}$, $6\sqrt{2}$, $6\sqrt{2}$.

10. (i) $a = 3$, $b = 7$, $c = -\frac{1}{3}$, (ii) $a = 3$, $c = -7$

(iii) $a = 2$, $b = 3$, $c = 0$ (iv) $a = 7$, $b = c$ (arbitrary)

12. (a) $\frac{9}{\sqrt{6}} (i + 2j + k)$ (b) $\frac{3}{\sqrt{11}} (i - j - 3k)$

14. (a) $20(1 + \sqrt{2})i - 20\sqrt{2}j$ (b) $-20(1 + \sqrt{2})i + 20\sqrt{2}j$

16. $3i + 2j + 3k$, $\sqrt{22}$ units

18. (a) $\left[\sqrt{2}, \frac{\pi}{4}, 1 \right]$ (b) $\left[\sqrt{3}, \frac{\pi}{4}, \arccos \left[\frac{1}{\sqrt{3}} \right] \right]$

20. (a) $\left[\sqrt{2}, -\frac{3\pi}{4}, -1 \right]$ (b) $\left[\sqrt{3}, -\frac{3\pi}{4}, \arccos \left[\frac{-1}{\sqrt{3}} \right] \right]$

22. $\left[\sqrt{27}, -\frac{\pi}{6}, 3 \right]$

Section 2.5

2. (i) 4 (ii) 51.9° (iii) $4/\sqrt{3}$ (iv) $4/\sqrt{14}$

4. (i) -100 (ii) 180° (iii) $\sqrt{50}$ (iv) $2\sqrt{50}$

10. (ii) and (iii) 16. $W = F.(b - a)$, $W = (25/\sqrt{3})J$

18. (a) $\alpha = \beta = \gamma = \arccos \frac{1}{\sqrt{3}} = 54.7^{\circ}$

(b) $\alpha = \arccos \left[\frac{1}{\sqrt{6}} \right] = 65.9^{\circ}$, $\beta = \arccos \left[\frac{-1}{\sqrt{6}} \right] = 114.1^{\circ}$,

$\gamma = \arccos \left[\frac{2}{\sqrt{6}} \right] = 35.3^{\circ}$.

(c) $\alpha = \arccos \left[\frac{3}{\sqrt{14}} \right] = 36.7^{\circ}$, $\beta = \arccos \left[\frac{1}{\sqrt{14}} \right] = 74.5^{\circ}$,

$\gamma = \arccos \left[\frac{2}{\sqrt{14}} \right] = 57.7^{\circ}$.

(d) $\alpha = \arccos \left[\frac{1}{2} \right] = 60^{\circ}$, $\beta = \arccos 0 = 90^{\circ}$,

$\gamma = \arccos \left[\frac{-\sqrt{3}}{2} \right] = 150^{\circ}$.

8

20. Resultant $R = 5i + j + 6k$ units. Equilibrium obtained by applying a force $F = -R = -5i - j - 6k$ units. $\|F\| = \sqrt{62}$ units, with direction cosines $\quad l = \dfrac{-5}{\sqrt{62}}$, $\quad m = \dfrac{-1}{\sqrt{62}}$, $\quad n = \dfrac{-6}{\sqrt{62}}$.

Section 2.6

2. $3i + 4j - 2k$, $-6i - 8j + 4k$, $\sqrt{29}$, $2\sqrt{29}$

4. $\sqrt{38}$ sq. units
6. $\sqrt{81}$ sq. units

8. $3i - 11j - 8k$, $3i + j + 4k$
10. $-3i + 2j - k$, 3

14. $\lambda(j + k)$ for any real $\lambda \neq 0$.

20. $M = -4i - 5j + 3k$ units

22. $24i + 3j - 9k$ units

Section 2.7

2. 2
4. -12

6. 0
8. 2

10. 4
22. 0, 0

24. $i + j$, $2j$,
26. $-i + j$, $-i + k$,

Section 2.8

2. $-i + 2j + 4k$ $(\lambda = -1)$, $i + j + 3k$ $(\lambda = 0)$, $3i + 2k$ $(\lambda = 1)$.

4. $r = 2i + 4j - k + \lambda(i + 2j + k)$

6. $r = 2j + \lambda(i + j + 2k)$
8. $r = i + 4j + k + \lambda(i - 6j + 2k)$

10. $r = i - j + 3k + \lambda(j - 2k)$

12. (i) $\dfrac{x - 3}{2} = \dfrac{y}{-1} = \dfrac{z - 1}{1}$

(ii) $\dfrac{x - 1}{1} = \dfrac{y + 2}{4} = \dfrac{z + 3}{-1}$

(iii) $\dfrac{x + 1}{1} = \dfrac{y - 1}{0} = \dfrac{z + 3}{2}$

16. $1/\sqrt{14}$ 18. $1/\sqrt{2}$

20. $\mathbf{n.r} = 0$, $2x - y + 2z = 0$ 22. $\mathbf{n.r} = 1$, $x + y + 2z = 1$

28. $\sqrt{3}$ 30. $\mathbf{r} = \frac{7}{5}\mathbf{i} + \frac{4}{5}\mathbf{j} + \lambda(\mathbf{j} - \mathbf{k})$

32. $\mathbf{r} = \frac{3}{2}\mathbf{i} + \frac{1}{2}\mathbf{j} + \lambda(\mathbf{j} + \mathbf{k})$

Section 2.9

2. $\mathbf{i} + \mathbf{j}$, $2\mathbf{i} + \mathbf{k}$; 2   4. $\sinh x$, $\cosh x$; 2

6. 1, x, x^2, x^3; 4   8. Infinite dimensional

10. No; zero is excluded so no closure under addition.

12. No; there is no closure because zero and negative integers are omitted.

14. No; because no closure under multiplication.

16. Yes; infinite dimensional.

18. No; no closure under multiplication by a scalar which could be irrational.

20. Yes;

$$(x_1, \tfrac{3}{4} x_1, 0, 0, ..., 0), (0, 0, 1, 0, ..., 0)$$

$$(0, 0, 0, 1, 0, ..., 0), ..., (0, 0, 0, 0, ..., 0, 1)$$

22. No; no closure under addition as zero is omitted.

24. $(\alpha, \beta, \gamma, 1, 0, ..., 0), (0, 0, 0, 0, 1, 0, ..., 0)$

, ..., $(0, 0, 0, ..., 0, 1)$. $((n - 1)$ vectors in all$)$

Chapter 3. Matrices.

Section 3.2

2. $\begin{bmatrix} -3 & 4 & -2 \\ -2 & 1 & -2 \\ 3 & 1 & -3 \end{bmatrix}$

4. $\begin{bmatrix} -5 & -6 & -8 \\ -6 & 1 & -10 \\ 1 & 3 & -3 \end{bmatrix}$

6. $\begin{bmatrix} 2 & 1 & 4 \\ 1 & 0 & 3 \\ 4 & 3 & 0 \end{bmatrix}$

8. $\begin{bmatrix} -1 & -1 & 7 \\ 5 & 1 & 4 \\ 2 & 1 & -3 \end{bmatrix}$

10. C^T

12. $S = \begin{bmatrix} 1 & 1/2 & 2 \\ 1/2 & 0 & 3/2 \\ 2 & 3/2 & 0 \end{bmatrix}$, $K = \begin{bmatrix} 0 & -3/2 & 1 \\ 3/2 & 0 & 5/2 \\ -1 & -5/2 & 0 \end{bmatrix}$

14. $S = \begin{bmatrix} -2 & 3/2 & 5/2 \\ 3/2 & 1 & 1 \\ 5/2 & 1 & -3 \end{bmatrix}$, $K = \begin{bmatrix} 0 & -3/2 & 3/2 \\ 3/2 & 0 & 1 \\ -3/2 & -1 & 0 \end{bmatrix}$

Use $S^T = S$ when S is symmetric and $K^T = -K$ when K is skew symmetric.

16. Yes. $\begin{bmatrix} 0 & 1 & 0 \\ -1 & 0 & 0 \\ 0 & 0 & 0 \end{bmatrix}$, $\begin{bmatrix} 0 & 0 & 1 \\ 0 & 0 & 0 \\ -1 & 0 & 0 \end{bmatrix}$, $\begin{bmatrix} 0 & 0 & 0 \\ 0 & 0 & 1 \\ 0 & -1 & 0 \end{bmatrix}$, 3

18. Each of the matrices B^T_{pq} is contained in the set of matrices B_{pq}, and conversely.

20. $\frac{1}{2}n(n-1)$. $\begin{bmatrix} 0 & 0 & 0 \\ 1 & 0 & 0 \\ 0 & 0 & 0 \end{bmatrix}$, $\begin{bmatrix} 0 & 0 & 0 \\ 0 & 0 & 0 \\ 1 & 0 & 0 \end{bmatrix}$, $\begin{bmatrix} 0 & 0 & 0 \\ 0 & 0 & 0 \\ 0 & 1 & 0 \end{bmatrix}$

22.

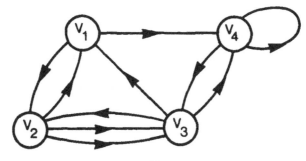

11

30. $\mathbf{P} = \begin{bmatrix} 0 & 0 & 0 & 0 & 1 \\ 0 & 0 & 0 & 1 & 0 \\ p_2 & p_3 & 0 & 0 & p_1 \\ 0 & q_2 & q_3 & 0 & q_1 \\ 0 & 0 & r_2 & r_3 & r_1 \end{bmatrix}$

Section 3.3

2. $\begin{bmatrix} 3 & -3 & 6 \\ 2 & -2 & 4 \\ 1 & -1 & 2 \end{bmatrix}$
            4.   Not defined

6. $\begin{bmatrix} 9 \\ 0 \end{bmatrix}$
            8. $[8 \quad -4 \quad 19]$

10. $\begin{bmatrix} 6 & 1 \\ 1 & 6 \end{bmatrix}$
             12. $\begin{bmatrix} 21 & 14 & 1 \\ 14 & 13 & 5 \\ 1 & 5 & 6 \end{bmatrix}$

14. $\mathbf{BA} = \begin{bmatrix} 5 & 6 & 2 \\ 7 & 6 & 4 \\ 5 & 5 & 2 \end{bmatrix}$, $\mathbf{AB} = \begin{bmatrix} 4 & 5 & 5 \\ 3 & 3 & 3 \\ 7 & 8 & 6 \end{bmatrix}$

18. (i) $\mathbf{ABAB}$ (ii) $\mathbf{B^T A^T B^T A^T}$ (iii) $\mathbf{BA^T BA^T}$ (iv) $\mathbf{A^2 B^2}$

20. $\mathbf{y} = \mathbf{Cz}, \ \mathbf{y} = \begin{bmatrix} y_1 \\ y_2 \end{bmatrix}$, $\mathbf{C} = \begin{bmatrix} 14 & -1 \\ 7 & 4 \end{bmatrix}$, $\mathbf{z} = \begin{bmatrix} z_1 \\ z_2 \end{bmatrix}$

   $y_1 = 14z_1 - z_2$, $y_2 = 7z_1 + 4z_2$.

22. (a) Result follows directly because rows of $\mathbf{A}$ are columns of $\mathbf{A}^T$.

   (b) (i) No   (ii) Yes   (iii) Yes   (iv) Yes   (v) No   (vi) Yes

28. $\mathbf{a}_{n+1} = \mathbf{A}^{n+1} \mathbf{a}.$

32. The result for A^n follows by computing A^2 and then using induction to establish the stated property (or by inspection having computed A^2, A^3, A^4 and used the formula for the sum of the first n integers which form an arithmetic series). The last result follows by summing terms in each position of the matrix and using the series for e^t.

34. (a) $A^2 = \begin{bmatrix} -(b^2 + c^2) & ab & ac \\ ab & -(c^2 + a^2) & cb \\ ca & cb & -(a^2 + b^2) \end{bmatrix} =$

$$\begin{bmatrix} -1 + a^2 & ab & ac \\ ab & -1 + b^2 & bc \\ ca & cb & -1 + c^2 \end{bmatrix} = -I + \begin{bmatrix} a^2 & ab & ac \\ ab & b^2 & bc \\ ac & bc & c^2 \end{bmatrix} = -I + x^T x .$$

(b) $A^2 = -I + x^T x$, so $A^3 = -A + x^T x A$

$$= -A + x^T(xA) = -A + 0 = -A .$$

(c) $A^3 = -A$, so $A^4 = -A^2 = \begin{bmatrix} -a^2+1 & -ab & -ac \\ -ab & -b^2+1 & -bc \\ -ac & -bc & -c^2+1 \end{bmatrix} .$

Section 3.4

2. $x_1 = 1$, $x_2 = 2$, $x_3 = 6$ 4. $x_1 = 3$, $x_2 = -1$, $x_3 = 2$

6. $x_1 = k$, $x_2 = 2k$, $x_3 = 3k$ (k arbitrary)

8. Only the trivial solution

10. $x_1 = k$, $x_2 = 2-3k$, $x_3 = 1+k$ (k arbitrary)

12. $x_1 = -k$, $x_2 = k+1$, $x_3 = -k$, $x_4 = 2k+3$ (k arbitrary)

14. $x_1 = 7-13k$, $x_2 = 7k-\ell-3$, $x_3 = k$, $x_4 = \ell$ (k, ℓ arbitrary)

16. $x_1 = -8k-\ell$, $x_2 = 1+2k+\ell$, $x_3 = k$, $x_4 = \ell$ (k, ℓ arbitrary)

18. $I_{AB} = \dfrac{E_0}{2R}$, $I_{AC} = \dfrac{E}{2R}$, $I_{AD} = \dfrac{E}{R}$, $I_{BD} = \dfrac{E_0}{2R}$, $I_{BC} = 0$

Since $I_{BC} = 0$ the value of the resistor BC is immaterial.

20. $I_1 = \dfrac{E_1 - E_2}{R_1}$, $I_2 = E_2(\dfrac{1}{R_1} + \dfrac{1}{R_2}) - \dfrac{E_1}{R_1}$, $I_3 = \dfrac{E_2}{R_2}$

22. $\dfrac{dx}{dt} = Ax - b$

$$x = \begin{bmatrix} x_1 \\ x_2 \\ x_3 \end{bmatrix}, \quad \frac{dx}{dt}\begin{bmatrix} \dfrac{dx_1}{dt} \\ \dfrac{dx_2}{dt} \\ \dfrac{dx_3}{dt} \end{bmatrix}, \quad A = \begin{bmatrix} 6 & -1 & 0 \\ 0 & 3 & 2 \\ 0.1 & 2 & -1 \end{bmatrix}, \quad b = \begin{bmatrix} 59.5 \\ 4.5 \\ 0.5 \end{bmatrix}$$

Equilibrium solution $x_1 = 10$, $x_2 = 0.5$, $x_3 = 1.5$ corresponds to $\dfrac{dx}{dt} = 0$.

Section 3.5

2. Linearly dependent

4. Linearly dependent

6. Linearly independent

8. 1

10. 2 ($R_3 = R_2 - R_1$)

12. 3 (consider $R_1 - 3R_4$, $R_2 - R_3 - 3R_4$, $R_3 - R_4$, $R_4 = R_4$)

14. Take $A = \begin{bmatrix} 1 & 0 \\ 1 & 0 \end{bmatrix}$, $B = \begin{bmatrix} 1 & 1 \\ 1 & 2 \end{bmatrix}$, then rank $A = 1$, rank $B = 1$ and

rank $(AB) = 1$

16. $A_E = \begin{bmatrix} 1 & 0 & 0 & \frac{1}{2} \\ 0 & 1 & \frac{1}{2} & \frac{1}{4} \\ 0 & 0 & 0 & 0 \end{bmatrix}$

(i) rank $A = 2$ (ii) $[1, 0, 0, \frac{1}{2}]$, $[0, 1, \frac{1}{2}, \frac{1}{4}]$ (iii) $\begin{bmatrix} 1 \\ 0 \\ 0 \end{bmatrix}$, $\begin{bmatrix} 0 \\ 1 \\ 0 \end{bmatrix}$

18. $A_E = \begin{bmatrix} 1 & 0 & 0 & 0 \\ 0 & 1 & 0 & 0 \\ 0 & 0 & 1 & 0 \\ 0 & 0 & 0 & 1 \\ 0 & 0 & 0 & 0 \end{bmatrix}$

(i) rank A = 4   (ii) [1, 0, 0, 0], [0, 1, 0, 0], [0, 0, 1, 0], [0, 0, 0, 1]

(iii) $\begin{bmatrix} 1 \\ 0 \\ 0 \\ 0 \\ 0 \end{bmatrix}, \begin{bmatrix} 0 \\ 1 \\ 0 \\ 0 \\ 0 \end{bmatrix}, \begin{bmatrix} 0 \\ 0 \\ 1 \\ 0 \\ 0 \end{bmatrix}, \begin{bmatrix} 0 \\ 0 \\ 0 \\ 1 \\ 0 \end{bmatrix}$

20. $A_E = \begin{bmatrix} 1 & 0 & 0 & 3 & 2 & 4 \\ 0 & 1 & 0 & 2 & 3 & -1 \\ 0 & 0 & 1 & -4 & -5 & 0 \end{bmatrix}$

(i) rank A = 3   (ii) [1, 0, 0, 3, 2, 4], [0, 1, 0, 2, 3, −1],

[0, 0, 1, −4, −5, 0]   (iii) $\begin{bmatrix} 1 \\ 0 \\ 0 \end{bmatrix}, \begin{bmatrix} 0 \\ 1 \\ 0 \end{bmatrix}, \begin{bmatrix} 0 \\ 0 \\ 1 \end{bmatrix}$

Section 3.7

2. (i) 24   (ii) abc   4.   (i) 0   (ii) −28

6. −34 in each case

8. −70 (interchange C_2 and C_3 and remove factor −2 from new C_2)

10. −35 (replace C_1 by $C_1 + 4C_3$)

12. −70 (remove factor 2 from R_1)

20. $C_1 + C_2 - C_3$, remove factor 2, $C_2 - C_1$, $C_3 - C_2$

24. 0.   Result also follows without differentiation because first two rows are linearly dependent so determinant is zero.

Section 3.8

2. $x_1 = -1$, $x_2 = 2$

4. $D = D_1 = D_2 = 0$; $x_1 = k$, $x_2 = \frac{1}{2} + \frac{7}{4}k$ (k arbitrary)

6. $x_1 = 2$, $x_2 = -2$, $x_3 = -1$ 8. $x_1 = x_2 = x_3 = x_4 = 0$

10. $x_1 = 1$, $x_2 = 0$, $x_3 = -1$, $x_4 = -2$

Section 3.9

2. No solution needed

4. $\det A = 0$, no inverse

6. $\begin{bmatrix} \cos\theta & \sin\theta \\ \sin\theta & -\sin\theta \end{bmatrix}$

8. $\begin{bmatrix} \frac{5}{16} & 0 & 0 \\ 0 & \frac{5}{2} & 0 \\ 0 & 0 & \frac{5}{8} \end{bmatrix}$

10. $\begin{bmatrix} \frac{3}{8} & -\frac{1}{4} & \frac{1}{8} \\ \frac{3}{16} & \frac{3}{8} & \frac{1}{16} \\ -\frac{1}{16} & -\frac{1}{8} & \frac{5}{16} \end{bmatrix}$

12. $x_1 = 1$, $x_2 = 2$, $x_3 = -3$

14. $x_1 = -2$, $x_2 = 1$, $x_3 = 4$

16. $x_1 = 2y_1 \qquad + \; y_2$
     $x_2 = 2y_2 \qquad + \; y_3$
     $x_3 = y_1 - y_2 + 4y_3$

24. $A^{-1} = \begin{bmatrix} 1 & 0 & 0 \\ 0 & \frac{1}{\sqrt{2}} & \frac{1}{\sqrt{2}} \\ 0 & -\frac{1}{\sqrt{2}} & \frac{1}{\sqrt{2}} \end{bmatrix}$

26. $A^{-1} = \frac{1}{11} \begin{bmatrix} 2 & 1 & 1 \\ 7 & -2 & -2 \\ 8 & 4 & -7 \end{bmatrix}$

28. A is singular ; no inverse

30. $x = \begin{bmatrix} 2 \\ -1.5 \\ 2.5 \end{bmatrix}$

32. $v = Zi$, so $i = Z^{-1}v$

34. $\det T = z_{12}^2/z_{12}^2 = 1$ so T is nonsingular and $x_2 = T^{-1}x_1$

36. $B^{-1} = \begin{bmatrix} 6 & -2 & -3 \\ -1 & 1 & 0 \\ -1 & 0 & 1 \end{bmatrix}$.

36. Cont'd

$A + I = B$, so $X = B^{-1}(A - I)$

giving $X = \begin{bmatrix} 11 & 4 & 6 \\ 2 & -1 & 0 \\ 2 & 0 & -1 \end{bmatrix}$.

Section 3.10

2. $\lambda_1 = 1$, $\lambda_2 = -11$, $\rho = 11$, $x_1 = \begin{bmatrix} \alpha \\ 0 \end{bmatrix}$, $x_2 = \begin{bmatrix} 0 \\ \beta \end{bmatrix}$, $\alpha, \beta \neq 0$ arbitrary

$\hat{x}_1 = \begin{bmatrix} 1 \\ 0 \end{bmatrix}$, $\hat{x}_2 = \begin{bmatrix} 0 \\ 1 \end{bmatrix}$, basis x_1 , x_2

4. $\lambda_1 = 1$, $\lambda_2 = 3$, $\rho = 3$, $x_1 = \alpha \begin{bmatrix} 1 \\ -1 \end{bmatrix}$, $x_2 = \beta \begin{bmatrix} 1 \\ 1 \end{bmatrix}$, $\alpha, \beta \neq 0$ arbitrary

$\hat{x}_1 = \frac{1}{\sqrt{2}} \begin{bmatrix} 1 \\ -1 \end{bmatrix}$, $\hat{x}_2 = \frac{1}{\sqrt{2}} \begin{bmatrix} 1 \\ 1 \end{bmatrix}$, basis x_1 , x_2

6. $\lambda_1 = 1$, $\lambda_2 = 6$, $\rho = 6$, $x_1 = \alpha \begin{bmatrix} 1 \\ -1 \end{bmatrix}$, $x_2 = \beta \begin{bmatrix} 4 \\ 1 \end{bmatrix}$, $\alpha, \beta \neq 0$ arbitrary

$\hat{x}_1 = \frac{1}{\sqrt{2}} \begin{bmatrix} 1 \\ -1 \end{bmatrix}$, $\hat{x}_2 = \frac{1}{\sqrt{17}} \begin{bmatrix} 4 \\ 1 \end{bmatrix}$, basis x_1 , x_2

8. $\lambda_1 = 0$, $\lambda_2 = -i$, $\lambda_3 = i$, $\rho = 1$, $x_1 = \alpha \begin{bmatrix} 1 \\ \frac{2}{3} \\ -\frac{1}{3} \end{bmatrix}$, $\alpha \neq 0$ arbitrary

$\hat{x}_1 = \frac{3}{\sqrt{14}} \begin{bmatrix} 1 \\ \frac{2}{3} \\ -\frac{1}{3} \end{bmatrix}$, x_2 , x_3 are complex conjugates; basis x_1 , x_2 , x_3 .

10. $\lambda_1 = 1$, $\lambda_2 = \lambda_3 = 7$ (algebraic multiplicity 2), $\rho = 7$, $x_1 = \begin{bmatrix} 0 \\ 0 \\ \alpha \end{bmatrix}$,

$x_2 = \begin{bmatrix} 0 \\ \beta \\ 0 \end{bmatrix}$, $\alpha, \beta \neq 0$ arbitrary ($\lambda = 7$ has geometric multiplicity 1) ;

basis x_1 , x_2

12. $\lambda_1 = -11$, $\lambda_2 = \lambda_3 = 2$ (algebraic multiplicity 2), $\rho = 11$,

$$x_1 = \begin{bmatrix} \alpha \\ 0 \\ 0 \end{bmatrix} ,$$

$$x_2 = \begin{bmatrix} 0 \\ \beta \\ 0 \end{bmatrix} , \ \alpha, \beta \neq 0 \text{ arbitrary } (\lambda = 2 \text{ has geometric multiplicity } 1)$$

$$\hat{x}_1 = \begin{bmatrix} 1 \\ 0 \\ 0 \end{bmatrix} , \ \hat{x}_2 = \begin{bmatrix} 0 \\ 1 \\ 0 \end{bmatrix} \quad \text{basis } x_1 , \ x_2$$

16. Eigenvalues of A, $\lambda_1 = -1$, $\lambda_2 = 6$. Eigenvectors of A and R(A)

$$x_1 = \begin{bmatrix} 2\alpha \\ -5\alpha \end{bmatrix}$$

$$x_2 = \begin{bmatrix} \beta \\ \beta \end{bmatrix} , \ \alpha, \beta \neq 0 \text{ arbitrary, } R(\lambda_1) = 9 , \ R(\lambda_2) = \frac{23}{71}$$

18. Eigenvalues of A, $\lambda_1 = 3$, $\lambda_2 = 9$. Eigenvectors of A and R(A)

$$x_1 = \begin{bmatrix} 4\alpha \\ \alpha \end{bmatrix}$$

$$x_2 = \begin{bmatrix} \beta \\ \beta \end{bmatrix} , \ \alpha, \beta \neq 0 \text{ arbitrary, } R(\lambda_1) = \frac{4}{5} , \ R(\lambda_2) = \frac{121}{122}$$

22. Eigenvalues -1, 1, 2.

24. Eigenvalues 1, 2 and 2.

C_1 centered at 2 , radius $\rho_1 = 0$

C_2 centered at 2 , radius $\rho_2 = \frac{1}{2}$

C_3 centered at 1 , radius $\rho_3 = \frac{1}{4}$

Eigenvalues and circles all lie to right of the imaginary axis.

26. Eigenvalues -3, -3 and 6.

C_1 centered at 1 , $\rho_1 = 6$

C_2 centered at -2 , $\rho_2 = 4$

26. Cont'd...

C$_3$ centered at 1 , $\rho_3 = 6$

Eigenvalues and circles lie on each side of the imaginary axis.

38. $\mathbf{k}_1 = \begin{bmatrix} 1 \\ -1 \end{bmatrix}$, $\mathbf{k}_2 = \begin{bmatrix} \frac{1}{2} \\ \frac{1}{2} \end{bmatrix}$, $Q = \begin{bmatrix} \frac{1}{\sqrt{2}} & \frac{1}{\sqrt{2}} \\ \frac{-1}{\sqrt{2}} & \frac{-1}{\sqrt{2}} \end{bmatrix}$

40. $\mathbf{k}_1 = \begin{bmatrix} 1 \\ 0 \\ 0 \end{bmatrix}$, $\mathbf{k}_2 = \begin{bmatrix} 0 \\ -1 \\ 0 \end{bmatrix}$, $\mathbf{k}_3 = \begin{bmatrix} 0 \\ 0 \\ 2 \end{bmatrix}$, $Q = \begin{bmatrix} 1 & 0 & 0 \\ 0 & 1 & 0 \\ 0 & 0 & 1 \end{bmatrix}$

Section 3.11

2. Eigenvalues −6, 3, 3, $P = \begin{bmatrix} -2 & 0 & 1 \\ 1 & -2 & 2 \\ 2 & 1 & 0 \end{bmatrix}$

4. Eigenvalues 2, 4, 4, $P = \begin{bmatrix} 0 & 1 & 0 \\ 1 & 0 & 1 \\ -1 & 0 & 1 \end{bmatrix}$

6. Eigenvalues −3, −3, 6, $P = \begin{bmatrix} -2 & 1 & 2 \\ 2 & 2 & 1 \\ 1 & -2 & 2 \end{bmatrix}$

8. Eigenvalues −2, −2, 4, $P = \begin{bmatrix} 0 & 1 & 1 \\ 1 & 0 & 0 \\ 0 & 1 & -1 \end{bmatrix}$

10. $A^{-1} = \frac{1}{12} \begin{bmatrix} -1 & 1 \\ 11 & 1 \end{bmatrix}$, $A^3 = \begin{bmatrix} -12 & 12 \\ 132 & 12 \end{bmatrix}$

12. $A^{-1} = \frac{1}{2} \begin{bmatrix} 0 & 1 & 1 \\ 2 & 0 & 0 \\ 0 & 1 & -1 \end{bmatrix}$, $A^4 = \begin{bmatrix} 0 & 1 & 2 \\ 3 & 0 & -1 \\ -1 & 2 & -1 \end{bmatrix}$

14. $\quad A^{-1} = \begin{bmatrix} 0 & 1 & -1 \\ 1 & -1 & 0 \\ 0 & 0 & 1 \end{bmatrix}$, $\quad A^4 = \begin{bmatrix} 5 & 3 & 11 \\ 3 & 2 & 7 \\ 0 & 0 & 1 \end{bmatrix}$

16. It follows from (15) that $c_0 A^{-1} = - [(-1)^n A^{n-1} + c_{n-1} A^{n-2}$

$+ \dots + c_2 A + c_1 I]$, so

$c_0 A^{-2} = [(-1)^n A^{n-2} + c_{n-1} A^{n-3} + \dots + c_2 I + c_1 A^{-1}]$.

Substitute for A^{-1} and the result follows.

18. Eigenvalues are 1, −1.

$$A^r = \left[\frac{1 - (-1)^r}{2}\right] A + \left[\frac{1 + (-1)^r}{2}\right] I \ .$$

The first result follows from the properties of the coefficients of A and I,

according as r is even or odd.

20. Eigenvalues are 1, 1, 2, so a double eigenvalue is involved.

$$A^r = (2^r - r - 1)A^2 + (2 + 3r - 2^{r+1})A + (2^r - 2r)I \ .$$

22. Eigenvalues are 1, 1, 1, so a triple eigenvalue is involved.

$$A^r = \tfrac{1}{2} r(r - 1)A^2 + (2r - r^2)A + (1 - \tfrac{3}{2} r + \tfrac{1}{2} r^2)I \ .$$

The final result follows by forming the matrix sum.

24. Eigenvalues are 2, 2, 1, so a double eigenvalue is involved.

$$A^r = (1 + r2^{r-1} - 2^r)A^2 + (2^{r+2} - 3r2^{r-1} - 4)A + (4 - 3.2^r$$
$$+ 2r2^{r-1})I \ .$$

26. Multiply by $A + I$ and then proceed as in Problem 25.

28. Multiply by $A + I$ and then proceed as in Problem 25.

30. Eigenvalues are i, $-i$, and the result for A^r, and hence for A^{2m} and

A^{2m-1}, follow in the usual manner. After rearrangement of terms we

find

$$\sum_{r=0}^{\infty} \frac{t^r}{r!} A^r = \sum_{m=0}^{\infty} \frac{t^{2m}}{(2m)!} A^{2m} + \sum_{m=1}^{\infty} \frac{t^{2m-1}}{(2m - 1)!} A^{2m-1}$$

30. Cont'd...

$$= \sum_{m=0}^{\infty} \frac{(-1)^m (kt)^{2m}}{(2m)!} I + \frac{1}{k} \sum_{m=1}^{\infty} \frac{(-1)^{m-1} (kt)^{2m-1}}{(2m-1)!} A$$

$$= \cos kt\ I + \frac{1}{k} \sin kt\ A \quad .$$

Rearrangement of terms is permissible because the series involved are absolutely convergent (see Sec. 8.1.).

Section 3.12

2. $A = \begin{bmatrix} 0 & -3 & 4 \\ 5 & 0 & -11 \\ 1 & -3 & 0 \end{bmatrix}$

   4. $A = \begin{bmatrix} 7 & 5 & 3 \\ 5 & 0 & -1 \\ 3 & -1 & 4 \end{bmatrix}$

8. $\Phi = -z_1^2 + z_2^2 + 2z_3^2$; indefinite

$x_1 = \frac{2}{3} z_1 + \frac{2}{3} z_2 - \frac{1}{3} z_3$, $x_2 = \frac{2}{3} z_1 - \frac{1}{3} z_2 + \frac{2}{3} z_3$,

$x_3 = -\frac{1}{3} z_1 + \frac{2}{3} z_2 + \frac{2}{3} z_3$;

$z_1 = \frac{2}{3} x_1 + \frac{2}{3} x_2 - \frac{1}{3} x_3$, $z_2 = \frac{2}{3} x_1 - \frac{1}{3} x_2 + \frac{2}{3} x_3$,

$z_3 = -\frac{1}{3} x_1 + \frac{2}{3} x_2 + \frac{2}{3} x_3$

10. $\Phi = -z_1^2 + 3z_3^2$; indefinite

$x_1 = \frac{3}{7} z_1 + \frac{6}{7} z_2 + \frac{2}{7} z_3$, $x_2 = \frac{6}{7} z_1 - \frac{2}{7} z_2 - \frac{3}{7} z_3$,

$x_3 = \frac{2}{7} z_1 - \frac{3}{7} z_2 + \frac{6}{7} z_3$;

$z_1 = \frac{3}{7} x_1 + \frac{6}{7} x_2 + \frac{2}{7} x_3$, $z_2 = \frac{6}{7} x_1 - \frac{2}{7} x_2 - \frac{3}{7} x_3$,

$z_3 = \frac{2}{7} x_1 - \frac{3}{7} x_2 + \frac{6}{7} x_3$

12. $\Phi = 2z_2^2 + 3z_3^2$; positive definite

$x_1 = \frac{1}{3} z_1 - \frac{2}{3} z_2 + \frac{2}{3} z_3$, $x_2 = -\frac{2}{3} z_1 + \frac{1}{3} z_2 + \frac{2}{3} z_3$,

$x_3 = -\frac{2}{3} z_1 - \frac{2}{3} z_2 - \frac{1}{3} z_3$;

12. Cont'd...

$$z_1 = \frac{1}{3} x_1 - \frac{2}{3} x_2 - \frac{2}{3} x_3 \quad , \quad z_2 = -\frac{2}{3} x_1 + \frac{1}{3} x_2 - \frac{2}{3} x_3 \quad ,$$

$$z_3 = \frac{2}{3} x_1 + \frac{2}{3} x_2 - \frac{1}{3} x_3$$

14. $A = \begin{bmatrix} \frac{-1}{3} & 0 & \frac{2}{3} \\ 0 & \frac{1}{3} & \frac{2}{3} \\ \frac{2}{3} & \frac{2}{3} & 0 \end{bmatrix}$ 16. Positive definite

18. Negative definite   20. Negative definite

22. The stationary points occur where $f_{x_1} = f_{x_2} = f_{x_3} = 0$; namely, at

$(\pm \, 1/\sqrt{6}, \, \pm \, 1/\sqrt{6}, \, \pm \, 1/\sqrt{6})$.

$$f_{x_1 x_1} = [4x_1^2(x_1 + x_2 + x_3) - 6x_1 - 2x_2 - 2x_3]e^{-(x_1^2 + x_2^2 + x_3^2)} \, ,$$

$$f_{x_1 x_2} = [4_{x_1 x_2}(x_1 + x_2 + x_3) - 2(x_1 + x_2)]e^{-(x_1^2 + x_2^2 + x_3^2)} \, ,$$

with corresponding expressions for the other derivatives obtained by

permuting $x_1, \, x_2, \, x_3$. Thus

$$H_{\pm} = \mp \, \frac{2}{\sqrt{6}} \, e^{-1/2} \begin{bmatrix} 4 & 1 & 1 \\ 1 & 4 & 1 \\ 1 & 1 & 4 \end{bmatrix}$$

Theorem 3.33(ii) shows H_+ is negative definite so $(1/\sqrt{6}, \, 1/\sqrt{6}, \, 1/\sqrt{6})$ is a

maximum, while H_- is positive definite so $(-1/\sqrt{6}, \, -1/\sqrt{6}, \, -1/\sqrt{6})$ is a

minimum.

Section 3.13

2. $L = \begin{bmatrix} 1 & 0 & 0 \\ 2 & 1 & 0 \\ 1 & -1 & 1 \end{bmatrix} , \quad U = \begin{bmatrix} 3 & -2 & 1 \\ 0 & 1 & 2 \\ 0 & 0 & 3 \end{bmatrix} .$

4. $L = \begin{bmatrix} 1 & 0 & 0 \\ -2 & 1 & 0 \\ -1 & 1 & 1 \end{bmatrix}$, $U = \begin{bmatrix} 5 & 1 & 2 \\ 0 & 1 & 1 \\ 0 & 0 & -1 \end{bmatrix}$; $x_1 = 2$, $x_2 = -2$, $x_3 = 1$

6. $L = \begin{bmatrix} 1 & 0 & 0 & 0 \\ 2 & 1 & 0 & 0 \\ 0 & -1 & 1 & 0 \\ 0 & 2 & 1 & 1 \end{bmatrix}$, $U = \begin{bmatrix} 3 & 0 & 1 & 0 \\ 0 & 1 & 2 & 0 \\ 0 & 0 & 1 & 4 \\ 0 & 0 & 0 & 1 \end{bmatrix}$; $x_1 = -1$, $x_2 = 1$, $x_3 = 2$, $x_4 = -2$.

8. $L = \begin{bmatrix} 1 & 0 & 0 & 0 \\ 1 & 1 & 0 & 0 \\ 1 & -1 & 1 & 0 \\ 0 & 0 & -3 & 1 \end{bmatrix}$, $U \begin{bmatrix} 1 & 1 & 0 & 1 \\ 0 & -1 & 0 & 1 \\ 0 & 0 & -1 & 0 \\ 0 & 0 & 0 & -1 \end{bmatrix}$; $x_1 = 2$, $x_2 = -1$, $x_3 = 1$, $x_4 = -1$

10. Positive definite; $Q = \begin{bmatrix} \sqrt{3} & 0 \\ -\sqrt{3} & \sqrt{6} \end{bmatrix}$.

12. Positive definite ; $Q = \begin{bmatrix} 1 & 0 & 0 \\ 0 & 1.2247 & 0 \\ 0 & -0.4083 & 1.1547 \end{bmatrix}$.

14. If all the elements d_1, d_2, ..., d_n of D are positive, setting $D = D_1^2$ with

$$D_1 = \begin{bmatrix} \sqrt{d_2} & 0 & . & . & . & 0 \\ 0 & \sqrt{d_2} & . & . & . & 0 \\ . & . & . & . & . & 0 \\ 0 & 0 & . & . & . & \sqrt{d_n} \end{bmatrix},$$

this becomes the Cholesky algorithm. A will be positive definite if all the $\sqrt{d_i}$ are positive, negative definite if all the $\sqrt{-d_i}$ are positive, and indefinite if at least two d_i are of opposite sign.

Chapter 4. First Order Ordinary Differential Equations.

Section 4.1

2. second order; particular solution

4. first order; general solution

6. fourth order; general solution

8. $y = A + 4x - \frac{1}{3} e^{-3x}$

10. $y = \frac{1}{60} x^5 + \frac{1}{2} Ax^2 + Bx + C$

12. $y = 2x + 2 \ln|x - 1| + A$

14. $y = \frac{1}{2} e^{2x} + \frac{2}{3} \cosh 3x + 2 - \frac{1}{2} e^2 - \frac{2}{3} \cosh 3$

16. $y = x(\ln|x - 2| - 1) - 2 \ln|x - 2| + 5x - 11$

18. $y = (x - 2)e^x + 5$

20. $(y')^2 = ky$

22. $L \frac{d^2q}{dt^2} + \frac{q}{C} = v(t)$

24. $\frac{dT}{dt} + \lambda T = \lambda h(t)$.

26. $\frac{dV}{dr} = 4\pi kr^2$ $(k > 0$ an arbitrary constant)

Section 4.2

2. Isoclines $x = k - 1$; exact solution $y = x + \frac{x^2}{2} + A$

4. Isoclines $x = \arcsin k$; exact solution $y = A - \cos x$

6. Isoclines $x = \ln(2x)$; exact solution $y = 2e^{\frac{1}{2}x} + A$

8. Isoclines $y = -\frac{1}{2} kx$

10. Isoclines $y = \pm\, x\sqrt{k/2}$ $(k \geq 0)$

12. Isoclines $y = \pm\, \sqrt{kx^2 - 1}$ $(x > \sqrt{1/k}$, $k > 0)$

14. Isoclines $x = k(1 + y^2)/y^2$

16. Isoclines $i = 2(\sin 2t - k)$

18. Isoclines $k = \frac{2}{\sqrt{\pi}} e^{-t^2}$

20. Isoclines $k = \sin\left[\frac{\pi}{2} t^2\right]$

Section 4.3

2. $y = c/x$

4. $y^2 = cx$

6. $y = x/(1 - cx)$

8. $cx = e^{(\frac{1}{2}y^2 + y)}$

10. $y = 2 + c \cos x$

12. $\ln|x| = c + \sqrt{y^2 + 1}$

14. $y^2 = 2(1 + e^{(1 + x)/x})$

16. $y \equiv 1$

18. $y = 3(x^2 - x)$

20. $\tan y = \dfrac{e^x - 1}{e - 1}$

22. $2y^2 \ln|cy| + x^2 = 0$

24. $c\sqrt{(x - 2)^2 + (y - 1)^2} = \exp\left\{\arctan\left[\dfrac{y - 1}{x - 2}\right]\right\}$.

26. $10y - 5x + 7 \ln|10x + 5y + 9| = c.$

28. $\ln|4x + 8y + 5| + 8y - 4x = c$

30. $\sqrt{4x + 2y - 1} - 2 \ln(\sqrt{4x + 2y - 1} + 2) = x + c$

32. $L\dfrac{di}{dt} + Ri = V_0$

34. If θ is the angle between OP and the x–axis, then $\theta = \arctan (y/x)$. Thus the angle between the tangent line at P and the x–axis is $2\theta = \arctan (y')$. Hence $2 \arctan (y/x) = \arctan (y')$. Taking the tangent of both sides of this equation and using the identity

$$\tan(A + B) = \frac{\tan A + \tan B}{1 - \tan A \tan B}$$

with $A = B = \arctan (y/x)$ gives the required result

$$y' = \frac{2xy}{x^2 - y^2} .$$

Integration of this homogeneous equation gives the general solution

$$y = c(x^2 + y^2),$$

so the solution through (x_0, y_0) is

$$y = \frac{y_0(x^2 + y^2)}{(x_0^2 + y_0^2)} .$$

36. Solution contains m/V lbs/gal of dissolved solid fertilizer at time t. As fluid is drawn off at a rate of Q gals/min, the rate of removal of dissolved solid (m/V)Q. Only water is replaced at the rate of Q gals/min, so

25

36. Cont'd...

equating the rate of change dm/dt to the removal rate gives

$$\frac{dm}{dt} = -\frac{m}{V} Q ,$$

where the negative sign is necessary because the dissolved mass is decreasing.

Section 4.4

2. $(y \sin x + xy \cos x + 2x)dx + (x \sin x - 2y)dy = 0$

4. $(\ln(1 + y^2) + 2xy^2)dx + \left[\dfrac{2xy}{1 + y^2} + 2x^2y\right]dy = 0$

6. $\left[\left[\sinh x + \dfrac{1}{y}\right] + 2\right]dx + \left[3y^2 - \dfrac{1}{y^2} \sinh\left[x + \dfrac{1}{y}\right]\right]dy = 0$

8. $(\tanh xy + xy \operatorname{sech}^2 xy)dx + x^2 \operatorname{sech}^2 xy\, dy = 0$

10. Exact; $x^2 + 4xy + y^2 + 8x + 2y = c$

12. Exact; $\ln|x^2 - 3y^2| + xy = c$

14. Exact; $\dfrac{x^2}{2} + xy + y^2 = c$

16. Exact; $\dfrac{x^4}{4} - \dfrac{3}{2} x^2y^2 + 2x + \dfrac{y^3}{3} = c$

18. Not exact; $\mu = 1/x^4$; $y^2 - x^2 = cx^3$

20. Not exact; $\mu = 1/x^2$; $\ln|x| - \dfrac{y^2}{x} = c$

22. Not exact; $\mu = 1/y^2$; $\dfrac{1}{y} \ln|x| + \dfrac{1}{2} y^2 = c$

24. Exact; $\dfrac{x^2}{2} + ye^{x/y} = \dfrac{1}{2} + 3e^{1/3}$.

26. Not exact; $\mu = 1/x^3$; $x - y^3 + 7x^2 = 0$

28. If $\mu = \mu(x)$ is an integrating factor

$$\frac{\partial}{\partial y} (\mu P) = \frac{\partial}{\partial x} (\mu Q) ,$$

and so

$$\mu \frac{\partial P}{\partial y} = \frac{d\mu}{dx} Q + \mu \frac{\partial Q}{\partial x} .$$

Differentiating partially with respect to y gives

28. Cont'd...

$$\mu \frac{\partial^2 P}{\partial y^2} = \frac{d\mu}{dx} \frac{\partial Q}{\partial y} + \mu \frac{\partial^2 Q}{\partial x \partial y} \ ,$$

because P and Q have continuous second order partial derivatives. Provided $Q \neq 0$, elimination of $d\mu/dx$ is possible between these last two equations, and it yields the required result. If $\mu = \mu(y)$ the condition becomes $P \neq 0$ and

$$P \left[\frac{\partial^2 P}{\partial x \partial y} - \frac{\partial^2 Q}{\partial x^2} \right] = \frac{\partial P}{\partial x} \left[\frac{\partial P}{\partial y} - \frac{\partial Q}{\partial x} \right] \ .$$

This result follows either by using a similar argument, or from symmetry by interchanging x and y, and P and Q.

Section 4.5

2. Subtracting c_2 times $y_2' + p(x)y_2 = q(x)$ from c_1 times $y_1' + p(x)y_1 = q(x)$ gives $(c_1 y_1 - c_2 y_2) + p(x)(c_1 y_1 - c_2 y_2) = 0$, showing $y = c_1 y_1 - c_2 y_2$ is a solution.

4. The expression $e^{\int p(x)dx}$ in (11) is an integrating factor, and if it is replaced by $e^{\int p(x)dx + A}$ the factor is multiplied by the arbitrary constant e^A. Similarly, the expression $e^{-\int p(x)dx}$ in (11) is the reciprocal of this integrating factor, and so if the constant A is included it is multiplied by the constant e^{-A}. When combined in (11) the product of the constants $e^A e^{-A} = 1$ is independent of A. Thus y_p contains no arbitrary constant irrespective of the choice of A.

6. $y = ce^{-x} + x(2 - x)$.

8. $y = ce^{-x} + x^2 - 2x + 4$.

10. $y = c(x + 1)^2 + (x + 1)^4$.

12. $y = ce^{-\sin x} + \sin x - 1$

14. $y = ce^{x^2} = x^2 - 1.$

16. $y = cx + x \ln(\ln|x|).$

18. $y = \frac{1}{x^3} (1 + \frac{1}{2} \ln|x|)^2)$.

20. $y = x \sec x$.

27

22. $y = \dfrac{(1 - a - a^2)}{a(1 - a)} \, x^a + \dfrac{x}{1 - a} - \dfrac{1}{a}$.

24. $x = c \sin y - \sin y \cos y$.

26. $x = 1/(3 + e^{-\cos y})$.

28. $x = \frac{1}{2}(y^5 - y^3)$.

30. $y = \dfrac{x}{c - 4 \ln |x|}$.

32. $y^2 = \dfrac{1}{(ce^{x^2} + 1 + x^2)}$.

34. $y^3 + 9x^3 \ln|x| = 8x^3$.

36. $y = \dfrac{1}{(1 + a \sqrt{1 - x^2} - a}$.

38. $x\left[c + (2 - y^2)e^{y^2/2}\right] = e^{y^2/2}$.

40. $T = (T_1 - T_0)e^{-\lambda t} + T_0$.

42. $i = \frac{1}{17} \left[8e^{-t/2} + 2 \sin 2t - 8 \cos 2t\right]$.

44. $P = \dfrac{1}{1 + \left[\dfrac{L}{P_0} - 1\right]e^{-at}}$.

46. $v = \left[U + \frac{g}{k}\right]e^{-kt} - \frac{g}{k}$.

48. A routine calculation establishes that

$$y = \dfrac{\omega b}{a^2 + \omega^2} e^{-at} + \dfrac{b}{a^2 + \omega^2} (a \sin \omega t - \omega \cos \omega t) .$$

Setting $a \sin \omega t - \omega \cos \omega t = R \sin (\omega - \varepsilon) = R(\sin \omega \cos \varepsilon$

$- \cos \omega \sin \varepsilon)$ it follows that

$$a = R \cos \varepsilon \text{ and } \omega = R \sin \varepsilon .$$

Thus

$$R = (a^2 + \omega^2)^{1/2}, \ \sin \varepsilon = \dfrac{\omega}{(a^2 + \omega^2)^{1/2}} , \ \cos \varepsilon = \dfrac{a}{(a^2 + \omega^2)^{1/2}}$$

from which the required result then follows at once.

50. $y_1 = x + \dfrac{1}{x}$, for $1 \le x \le 2$, $y_2 = \frac{1}{2}(8 - 3e^{(1-\frac{1}{2}x)})$, for $x \ge 2$. Yes.

52. $y_1 = (x + 3)e^{-x}$, for $0 \le x \le 1$, $y_2 = \frac{1}{3}(13e^{(2x-3)} - e^{-x})$, for

$x \ge 1$. No.

54. Results follow directly from the indicated substitutions. y is the general solution because it contains the necessary single arbitrary constant introuced when integrating either of the differential equations.

56. $y = \dfrac{c + x^2}{x(c - x^2)}$

58. $y = \dfrac{cx - 1 - \frac{1}{2}x \ln |x|}{cx - \frac{1}{2}x \ln|x|}$.

Section 4.6

2. radial lines $y = ax$.

4. $x^2 + \alpha y^2 = a$, corresponding to confocal ellipses when $\alpha > 0$ and rectangular hyperbolas when $\alpha < 0$.

6. coaxial circles $(x + a)^2 + y^2 = a^2 - 1$, $|a| \geq 1$.

8. $x^2 + y^2 - \ln y^2 = a$.

10. $x^2 + y^2 = \exp[2\sqrt{3} \text{ arc tan } (y/x)]$.

12. $y^2 - xy + 2x^2 = a \exp\left[\dfrac{6}{\sqrt{7}} \text{ arc tan }\left[\dfrac{2y - x}{\sqrt{7}}\right]\right]$.

14. cardioids $r = a(1 - \cos \theta)$, $a > 0$.

16. $r^m = a^m \sin m\theta$.

18. Let the angle between the tangent to a curve and its radius vector at a point be Φ. Let the angle between the curve and its isogonal trajectory be α, and let Φ' be the angle between the isogonal trajectory and the radius vector at the point. Then $\alpha = \Phi' - \Phi$. Thus if $k = \tan \alpha$,

$$k = \tan \alpha = \tan (\Phi' - \Phi) = \frac{\tan \Phi' - \tan \Phi}{1 + \tan \Phi' \tan \Phi} \ .$$

From (13) $\tan \Phi = r \dfrac{d\theta}{dr}$, so

$$\tan \Phi' = \frac{k + r(d\theta/dr)}{1 - kr(d\theta/dr)} \ .$$

Thus the required differential equation for the isogonal trajectories follows by replacing $r(d\theta/dr)$ in (15) by $\tan \Phi'$, and eliminating c between the resulting equation and $F(r, \theta, c) = 0$.

Section 4.7

2. $y \equiv 0$ is a solution and another is

$$y_a(x) = \begin{cases} 0, & \text{for } -\infty < x \leq a, \\ \frac{1}{3}(x - a)^3, & \text{for } x \geq a \text{ with } a \geq 0 \end{cases} \ .$$

4. The result follows directly from Theorem 4.3.

29

6. $|f| = |x + \sin y| \le 2$, so condition (i) of Theorem 4.3 is satisfied.

$|f(x, y_2) - f(x, y_1)| = |\sin y_2 - \sin y_1| = |2 \sin \frac{1}{2} (y_2 - y_1) \cos \frac{1}{2}$

$(y_2 + y_1)|$ so $|[f(x,y_2) - f(x,y_1)]/(y_2 - y_1)| \le 2|\cos \frac{1}{2} (y_2 + y_1)| \le 2$,

which shows condition (ii) of Theorem 4.3 is satisfied, and thus the solution

is unique.

8. $|f(x, y_2) - f(x, y_1)| = |x|y_2| - x|y_1|| \le |x||y_2 - y_1| <$

$a|y_2 - y_1|$

which shows f is Lipschitz continuous in the given region. The non–

existence of $\partial f/\partial y$ follows directly from its definition in terms of a limit.

The solution is unique in R because conditions (i) and (ii) of Theorem 4.3

are satisfied.

10. f(x, y) will be continuous in y alone in a region R if for $\varepsilon > 0$ and any

two points (x, y_1) and (x, y_2) in R there is a $\delta > 0$ such that

$$|f(x, y_2) - f(x, y_1)| < \varepsilon$$

when $|y_2 - y_1| < \delta$. The result follows by supposing f(x, y) satisfies a

Lipschitz condition in R with constant m and, given $\varepsilon > 0$, letting $\delta =$

ε/m.

12. $y_3(x) = \frac{39}{630} + \frac{49}{9} x - \frac{28}{3} x^2 + \frac{17}{3} x^3 - \frac{7}{18} x^4 + \frac{8}{15} x^5 + \frac{1}{63} x^7$.

14. $y_1(x) = x + \cos x$, $y_2(x) = \frac{x^2}{2} + \sin x + \cos x$,

$y_3(x) = 1 + \frac{x^3}{6} + \sin x$.

16. (a) $y_3(x) = 1 + x + x^2 + \frac{x^3}{3} + \frac{x^4}{24}$

(b) $y_3(x) = 1 + 2x + x^2 + \frac{x^3}{6} + \frac{x^4}{24} - \sin x$

(c) $y_3(x) = 1 + x^2 + \frac{x^3}{6} + \frac{x^4}{24} + \sinh x$.

Section 4.8

2.

n	x_n	k_{1n}	k_{2n}	k_{3n}	k_{4n}
0	0	0	0.013203	0.009316	0.044389
1	0.3	0.041276	0.082990	0.072105	0.118767
2	0.6	0.115753	0.153571	0.145143	0.180052
3	0.9	0.178519	0.204378	0.199442	0.221535
4	1.2	0.220882	0.237075	0.234394	0.247837
5	1.5	0.247564	0.257571	0.256116	0.264372

n	x_n	k_n	y_n
0	0	0.014905	1
1	0.3	0.078372	1.014905
2	0.6	0.148872	1.093277
3	0.9	0.201282	1.242149
4	1.2	0.235276	1.443431
5	1.5	0.256552	1.678707

4.

n	x_n	k_{1n}	k_{2n}	k_{3n}	k_{4n}
0	0	−0.3	−0.123961	−0.216311	−0.038074
1	0.3	−0.079421	−0.011811	−0.039049	0.023838
2	0.6	0.014807	0.047176	0.036990	0.065541
3	0.9	0.063049	0.080518	0.076124	0.091152
4	1.2	0.090340	0.100586	0.098471	0.107231
5	1.5	0.106925	0.113347	0.112231	0.117741

n	x_n	k_n	y_n
0	0	−0.169770	1
1	0.3	−0.026217	0.830230
2	0.6	0.041446	0.804013
3	0.9	0.077914	0.845459
4	1.2	0.099281	0.923374
5	1.5	0.112637	1.022654

6.

n	x_n	k_{1n}	k_{2n}	k_{3n}	k_{4n}
0	−1.0	0.5	0.460000	0.466144	0.326168
1	−0.5	0.340575	0.142057	0.248008	0.017785
2	0	0.095304	0.059910	0.082394	0.086947
3	0.5	0.092200	0.130270	0.111520	0.159286
4	1.0	0.150152	0.175734	0.166570	0.190429
5	1.5	0.187772	0.201488	0.197758	0.209546

n	x_n	k_n	y_n
0	−1.0	0.446409	0
1	−0.5	0.189748	0.446409
2	0	0.077810	0.636157
3	0.5	0.122511	0.713967
4	1.0	0.170865	0.836478
5	1.5	0.199302	1.007343

8.

n	x_n	k_{1n}	k_{2n}	k_{3n}	k_{4n}
0	0	0.2	0.200975	0.200985	0.203645
1	0.2	0.203656	0.207191	0.207298	0.210409
2	0.4	0.210399	0.211158	0.211197	0.207260
3	0.6	0.207161	0.195813	0.194996	0.173397
4	0.8	0.173185	0.139649	0.136662	0.090096
5	1.0	0.089967	0.033919	0.028430	−0.032044

n	x_n	k_n	y_n
0	0	0.201261	0
1	0.2	0.207174	0.201261
2	0.4	0.210395	0.408435
3	0.6	0.193696	0.618830
4	0.8	0.135984	0.812526
5	1.0	0.030437	0.948510

Chapter 5.   Linear Higher Order Differential Equations.

Section 5.1

12. $y = c_1 + c_2 x^2$ 14. $y = c_1 + c_2 \arctan x$.

16. $y = c_1 + c_2 \ln |x| + \frac{1}{9} x^3$.

20. $y = 2 - \sinh x + xe^{-x}$.

22. $y = (1 + x)e^x + 2$.   24. $y = c_1 - x + \ln|c_2 + e^{2x}|$.

26. $x = c_1 + k \ln \left| \sin \left[\frac{y - c_2}{k} \right] \right|$.

28. $\frac{1}{2} \ln|2y^3 + c_1| = x + c_2$.   30. $(x + c_1)^2 + (y + c_2)^2 = a^2$.

32. $(x + c_1)^2 + y^2 = c_2$ $(c_2 > 0)$.

34. Linearly independent : when expanded $(1 + x)^3$ is not expressible as a
     linear combination of x, x^2 and x^3 .

36. Linearly independent : $c_1 + c_2 \sinh^2 x + c_2 \cos^2 x = 0$ is only true for all
     x in $- 4 \le x \le 4$ if $c_1 = c_2 = c_3 = 0$.

40. Rearrange the equation to give

$$\phi_2' - \left[\frac{\phi_1'}{\phi_1} \right] \phi_2 = \left[A \; e^{-\int \frac{a_1(x)}{a_0(x)} \, dx} \right] / \phi_1 \; ,$$

and use the integrating factor $\mu = 1/\phi_1$.   The integration constant may be
omitted as it merely regenerates the term $k\phi_1$ in the general solution.

Section 5.2

6. $y = e^{x/4}(c_1 \cos \frac{\sqrt{7}}{4} x + c_2 \sin \frac{\sqrt{7}}{4} x)$.

8. $y = (c_1 + c_2 x)e^{2x}$.   10. $y = c_1 e^{-3x} + c_2 e^{\frac{1}{2}x}$.

12. $y = e^{-x}$.   14. $y = 1$.

16. $y = -3 \cos 4x$.   18. $y = -\frac{1}{2} e^{(\frac{\pi}{2} - x)} \sin 2x$.

20. $y'' - 2y' = 0$.   22. $y'' + 2y' + 3y = 0$.

24. $y'' - 9y = 0$.

26. $y = \dfrac{e^3}{(e^3 - 1)} (1 - e^{-3x})$.

28. $y = e^{(x-\pi/2)} \sin x$.

30. $y = (e^{(2+\sqrt{2})x} - e^{(2-\sqrt{2})x})/((2+\sqrt{2})e^{2+\sqrt{2}} - (2-\sqrt{2})e^{2-\sqrt{2}})$.

32. $k = \dfrac{(2n-1)\pi}{2a}$, with $n = 1, 2, ...$; $y = c \cos \left[\dfrac{(2n-1)\pi x}{2a}\right]$.

34. $k = n\pi/a$, with $n = 0, 1, 2, ...$; $y = c \sin \left[\dfrac{n\pi x}{a}\right]$.

36. $y = c_1 \cos kx + c_2 \sin kx$, so $y(0) + \pi y'(0) = 0$ shows $c_1 + \pi k c_2 = 0$, and $y(\pi) = 0$ shows $c_1 \cos k\pi + c_2 \sin k\pi = 0$. Thus unless $c_1 = c_2 = 0$ it follows that $\tan k\pi = k\pi$. The first two positive solutions are $k_1 = 1.430$, $k_2 = 2.459$. The number of solutions is infinite because of the infinite number of branches of the tangent function.

38. $y = 6e^{-2x}$.

40. $y = e^{(1+x)}$.

Section 5.3

2. linearly independent

4. linearly independent

6. linearly independent

8. $y = c_1 e^{3x} + c_2 e^{-2x}$; e^{3x}, e^{-2x} for all x .

10. $y = c_1 e^{-x} + c_2 x e^{-x} + c_3 e^{-3x}$; e^{-x}, $x e^{-x}$, e^{-3x} for all x .

12. $y = c_1 e^{-2x} \cos x + c_2 e^{-2x} \sin x$; $e^{-2x} \cos x$, $e^{-2x} \sin x$ for all x .

14. $y = c_1 e^{2x} + c_2 e^{-2x} + c_3 \cos 2x + c_3 \sin 2x$; e^{2x}, e^{-2x}, $\cos 2x$, $\sin 2x$ for all x .

16. $y = e^{x/\sqrt{2}}(c_1 \cos(x/\sqrt{2}) + c_2 \sin(x/\sqrt{2})) +$

$e^{-x/\sqrt{2}}(c_3 \cos(x/\sqrt{2}) + c_4 \sin(x/\sqrt{2}))$; $e^{x/\sqrt{2}} \cos(x/\sqrt{2})$,

$e^{x/\sqrt{2}} \sin(x/\sqrt{2})$, $e^{-x/\sqrt{2}} \cos(x/\sqrt{2})$, $e^{-x/\sqrt{2}} \sin(x/\sqrt{2})$ for all x .

18. $y = c_1e^{3x} + c_2xe^{3x} + c_3e^{-3x} + c_4xe^{-3x}$;

e^{3x}, xe^{3x}, e^{-3x}, xe^{-3x} .

20. $y = \frac{1}{2}(e^x + e^{-x}) = \cosh x$ 22. $y = \frac{1}{4}e^x + \frac{1}{2}(\frac{3}{2} + x)e^{-x}$.

24. $y = \frac{4}{3}\sin x - \frac{1}{6}\sin 2x$.

26. $y''' + y'' + 9y' + 9y = 0$; $y = c_1e^{-x} + c_2\cos 3x + c_3\sin 3x$;

$$y = \frac{1}{10}e^{(\frac{\pi}{2} - x)} - \frac{11}{30}\cos 3x + \frac{1}{10}\sin 3x .$$

Section 5.4

2. $-2(\sin x + 2\cos x)$, $8\cosh 3x$, $-4 - 3\sinh 2x$.

4. e^{-x}, $4 + 8(\sinh 2x + \cosh 2x)$, $2 + 8x + 4x^2 - 4\sin x + 3\cos x$.

6. $y = c_1e^{-x} + c_2e^x + c_3xe^x$.

8. $y = c_1e^{-x} + c_2xe^{-x} + c_3e^x + c_4xe^x$.

10. The solutions follow as a direct generalization of the argument for a double root given in Sec. 5.4. First set

$$P(\lambda) = (\lambda - \mu)^3 \, Q(\lambda) = 0$$

and by repeated differentiation with respect to λ show that $P(\mu) = P'(\mu) = P''(\mu) = 0$. Deduce the required results by starting from the fact that

$$P(D)[e^{\lambda x}] = P(\lambda)e^{\lambda x} = 0$$

and differentiating twice with respect to λ.

12. Setting $y = y_p = Ae^{\alpha x}$ in the differential operator $P(D)[y]$ we find $P(D)[Ae^{\alpha x}] = AP(\alpha)e^{\alpha x}$. If $P(D)[Ae^{\alpha x}] = ke^{\alpha x}$ it then follows that $A = k/P(\alpha)$, and hence that $y_p = [k/P(\alpha)]e^{\alpha x}$, because $P(\alpha) \neq 0$.

Section 5.5

2. $y = c_1e^x + c_2e^{7x} + 6/7$.   4. $y = (c_1 + c_2x)e^{-x} + \frac{1}{9}e^{2x}$.

6. $y = c_1 \cos x + c_2 \sin x - 2x \cos x$.

8. $y = e^{2x}(c_1 \cos x + c_2 \sin x) + \frac{1}{2} e^{2x} + \frac{1}{20} (2 \cos 2x + \sin 2x)$.

10. Use $\sin^2 x = (1 - \cos 2x)/2;$ $y = c_1 + c_2 e^{-x} + 5x + \frac{1}{10} (2 \cos 2x - \sin 2x)$.

12. $y = e^{2x}(c_1 \cos x + c_2 \sin x) - xe^{2x} \cos x$.

14. $y = e^x(c_1 \cos 3x + c_2 \sin 3x) + \frac{1}{37} (\sin 3x + 6 \cos 3x) + \frac{1}{5} e^{2x}$.

16. $y = c_1 + c_2 e^{3x} + c_3 e^{-3x} + 5x^2 - e^x$.

18. $y = c_1 + c_2 x + c_3 e^x + c_4 xe^x + \frac{1}{2} x^2 e^x + 12x^2 + 3x^3 + \frac{1}{2} x^4 + \frac{1}{20} x^5$.

20. $y = c_1 + c_2 x + c_3 x^2 + c_4 e^{-x} + \frac{1}{544} (4 \cos 4x - \sin 4x)$.

22. $y = a \cos nx + \dfrac{b(n^2 - p^2) - kp}{n(n^2 - p^2)} \sin nx + \dfrac{k}{n^2 - p^2} \sin px$.

24. $y = \frac{5}{8}\left[e^x + e^{-x}\right] - \frac{5}{4}\left[\cos x + x \sin x\right]$.

26. $y = e^{-2x}[(e^\pi - 1) \sin x - 2e^\pi \cos x - \cos 2x]$.

28. $y = c_1 e^x + c_2 e^{-x} - x \sin x - \cos x$.

30. $y = c_1 + c_2 e^{2x} - 2xe^x - \frac{3}{4}(x + x^2)$.

32. $y = c_1 e^{-x\sqrt{2}} + c_2 e^{x\sqrt{2}} + xe^x \sin x + e^x \cos x$.

34. $y = c_1 \cos x + c_2 \sin x - x \cos x + \sin x \ln |\sin x|$.

36. $y = c_1 e^x + c_2 xe^x + xe^x \ln |x|$.

38. $y = c_1 e^{-x} + c_2 xe^{-x} + xe^{-x} \ln |x|$.

40. $y = c_1 \cos x + c_2 \sin x + c_1 e^{2x} + c_2 e^{-2x} - \frac{1}{6} \cosh x$.

42. $y = c_1 + c_2 \sin x + c_3 \cos x + \ln |\sec x + \tan x| + \sin x \ln |\cos x| - x \cos x$.

46. $y = c_1 x^2 + \dfrac{c_2}{x}$.

48. $y = c_1 \cos (\ln|x|) + c_2 \sin (\ln|x|)$.

50. $y = c_1 \cos (\ln|x|) + c_2 \sin (\ln|x|) + 1$.

52. $y = \dfrac{c_1}{x} + c_2 x^2 + \dfrac{1}{10} (\cos (\ln|x|) - 3 \sin (\ln|x|))$.

Section 5.6

2. $y = c_1 + c_2 e^{-3x} + c_3 e^{3x}$.

4. $y = c_1 e^{-x} + c_2 e^{x} + c_3 \sin x + c_4 \cos x$.

6. $y = c_1 \sin (x^3) + c_2 \cos (x^3)$.

8. $y = c_1 + c_2 \ln(|x|) + c_3 x^4$.

Section 5.7

2. The tension in the string in equilibrium is $F = k\ell/4$. When the displacement is small the tension remains constant to terms of order x/ℓ. The total restoring force is $2F\cos \hat{APQ} = kx/2$ to terms of order x/ℓ. The equation of motion is thus

$$m \frac{d^2x}{dt^2} = - \frac{k}{2} x \ .$$

Angular frequency of free oscillations $\omega_0 = \sqrt{K/(2m)}$, so $T = 2\pi/\omega_0 = \pi\sqrt{8m/k}$.

4. As the center of mass remains fixed and the system is symmetric about the mass M the central mass must remain fixed and so will not enter into the equation of motion. The equation of motion then follows from Prob. 3 by setting $m_1 = m_2 = m$ to give

$$\frac{d^2x}{dt^2} = - \frac{2k}{m} x \ .$$

6. The restoring force $F = -kx$ and the viscous damping is $R = -\mu(dx/dt)$ so the equation of motion is thus

$$m \frac{d^2x}{dt^2} = -\mu \frac{dx}{dt} - kx .$$

8. Mass of fluid to be moved is $\rho A \ell$, so rate of change of momentum is $\rho A \ell (d^2x/dt^2)$. Hydrostatic pressure head restoring the level is $F = -2x\rho gA$, and the frictional resistance $R = -k (dx/dt)$. The equation of motion is thus

$$\rho A \ell \frac{d^2x}{dt^2} = -k\frac{dx}{dt} - 2x\rho gA .$$

If friction is neglected $k = 0$ and the angular frequency of the natural oscillations is $\omega_0 = \sqrt{2g/\ell}$. Thus $T = 2\pi/\omega_0 = \pi\sqrt{2\ell/g}$.

10. The rate of change of angular momentum is $I(d^2\theta/dt^2)$. As the angular displacement is small, the tension F remains constant to terms of order θ, so the restoring moment is approximately $M = -F\ell\theta/d$. Thus the equation of motion is

$$I \frac{d^2\theta}{dt^2} = -\frac{F\ell}{d} \theta .$$

The angular frequency of free oscillations is thus $\omega_0 = \sqrt{F\ell/(Id)}$ so the period of the oscillations $T = 2\pi/\omega_0 = 2\pi\sqrt{Id/(F\ell)}$.

12. $y = \exp(-\omega_0 t/(2Q)) \sin [\omega_0 t\sqrt{1 - \frac{1}{4Q^2}} + \varepsilon]$;

$y \simeq \exp(-\omega_0 t/(2Q)) \sin (\omega_0 t + \varepsilon)$.

18. $y_p = \frac{1}{3} \sin t$.   20. $y_p = \frac{1}{30} (2\cos t + \sin t)$.

22. $y_p = \frac{1}{13} (24\sin\frac{1}{2}t - 16\cos\frac{1}{2}t)$.

24. $y_c = c_1 \cos 2t + c_2 \sin 2t$.

26. $y_c = e^{-t}(c_1 \cos 2t + c_2 \sin 2t)$.

28. $y_c = e^{-\frac{1}{4}t}(c_1 \cos t + c_2 \sin t)$.

30. $y = 2 \sin 2t$.

32. $y = 3e^{-2t}(\cos \sqrt{2}\ t + \sqrt{2} \sin \sqrt{2}\ t)$.

34. $y = \cos 3t - \frac{5}{12} \sin 3t + \frac{1}{4} \sin t$.

36. $y = 2 \sin t + \frac{1}{2} t \sin t$.

38. $y = \dfrac{e^{-\frac{1}{3}t}}{411} (4 \cos \frac{\sqrt{2}}{3} t + 105\sqrt{2} \sin \frac{\sqrt{2}}{3}t) - \frac{1}{137} (4 \cos 2t + 11 \sin 2t)$.

40. $y_p = 2t \sin t$.   42. $y_p = \frac{1}{145} (18 \cos 2t + 16 \sin 2t)$.

Section 5.8

2. $v = e^{-\frac{1}{2}x}$; $u" - (25/4)u = 0$; $u_1 = e^{\frac{5x}{2}}$, $u_2 = e^{\frac{-5x}{2}}$;
$y = c_1 e^{2x} + c_2 e^{-3x}$

4. $v = e^{-3x}$; $u" = 0$; $u_1 = c_1$, $u_2 = c_2 x$; $y = (c_1 + c_2 x)e^{-3x}$.

6. $v = e^{2x}$; $u" = 0$; $u_1 = c_1$, $u_2 = c_2 x$; $y = (c_1 + c_2 x)e^{2x}$

8. $v = 1/x$; $u" + 9u = 0$; $u_1 = \cos 3x$, $u_2 = \sin 3x$;
$y = \frac{1}{x}(c_1 \cos 3x + c_2 \sin 3x)$

10. $v = e^{\frac{1}{2}x^2}$; $u" - (x^2 + 3)u = 0$.

12. $v = \bar{x}^{-1/2}$; $u" + \left[1 - \dfrac{15}{4x^2}\right]u = 0.$

14. $v = e^{-2x}$; $u" - \left[4 + \dfrac{2}{x}\right]u = 0$.

Section 5.9

2. $G(x,t) = \frac{1}{6}\left[e^{3(x - t)} - e^{-3(x - t)}\right]$;
$y = c_1 e^{-3x} + c_2 e^{3x} + \frac{1}{6} x e^{3x}$.

4. $G(x,t) = \sin(x - t)$;
$y = c_1 \cos x + c_2 \sin x + \frac{1}{2} x \sin x$.

6. $G(x,t) = \dfrac{x(x - t)}{t^3}$;

8. $G(x,t) = \begin{cases} -\dfrac{\sin 2x \cos 2(t-1)}{2 \cos 2} & , 0 \le x \le t \\[3mm] -\dfrac{\sin 2t \cos 2(x-1)}{2 \cos 2} & , t \le x \le 1 \end{cases}$;

$y = -\dfrac{1}{2 \cos 2}\left[\cos 2(x-1) \displaystyle\int_0^x \sin 2t f(t)dt + 2x\int_x^1 \cos 2(t-1)f(t)dt\right]$

10. $G(x,t) = \begin{cases} -x & , \quad 0 \le x \le t \\ -t & , \quad t \le x \le 1 \end{cases}$;

$y = \dfrac{1}{6}x^3 - \dfrac{1}{2}x$.

12. Integration gives $(1 - x^2)y' = c$, and thus the general solution

$y = \dfrac{c}{2}\left[\ln\left|\dfrac{1+x}{1-x}\right|\right] + d$.

Hence $\phi_1(x) = \ln\left|\dfrac{1+x}{1-x}\right|$ and $\phi_2(x) \equiv 1$.

$$G(x,t) = \begin{cases} -\dfrac{1}{2}\ln\left|\dfrac{1+x}{1-x}\right| & ,0 \le x \le t \\[3mm] -\dfrac{1}{2}\ln\left|\dfrac{1+t}{1-t}\right| & ,t \le x \le 1. \end{cases}$$

14. The result follows by showing that the given expression for y satisfies the differential equation. This is automatic for the complementary function, and it follows for the integral term by using Leibniz' theorem coupled with Theorem 3.18.

Chapter 6.   Systems of Linear Differential Equations.

Section 6.1

2. $C = \begin{bmatrix} 1 & 1 & 2 \\ 1 & -1 & 3 \\ 1 & 0 & 3 \end{bmatrix}$, $D = \begin{bmatrix} 1 & -1 & 1 \\ 1 & 2 & -1 \\ 4 & 2 & 1 \end{bmatrix}$, $x = \begin{bmatrix} x_1 \\ x_2 \\ x_3 \end{bmatrix}$.

4. $C = \begin{bmatrix} 1 & -2 & 3 \\ 0 & 1 & 4 \\ 0 & 0 & 7 \end{bmatrix}$, $D = \begin{bmatrix} 3 & 1 & -1 \\ 1 & 2 & 1 \\ 2 & 4 & -1 \end{bmatrix}$, $x = \begin{bmatrix} x_1 \\ x_2 \\ x_3 \end{bmatrix}$.

6. $A = \begin{bmatrix} 0 & 1 \\ -2 & -8/3 \end{bmatrix}$, $x = \begin{bmatrix} x_1 \\ x_2 \end{bmatrix}$

8. $A = \begin{bmatrix} 0 & 1 \\ 2/9 & 0 \end{bmatrix}$, $x = \begin{bmatrix} x_1 \\ x_2 \end{bmatrix}$.

10. $A = \begin{bmatrix} 0 & 1 & 0 \\ 0 & 0 & 1 \\ -4/3 & -1 & 3 \end{bmatrix}$, $x = \begin{bmatrix} x_1 \\ x_2 \\ x_3 \end{bmatrix}$.

12. $A = \begin{bmatrix} 1 & 1/10 \\ 0 & 3/10 \end{bmatrix}$, $x = \begin{bmatrix} x_1 \\ x_2 \end{bmatrix}$

14. $A = \begin{bmatrix} 5/6 & 2/3 & -2 \\ 1/3 & 2/3 & -1 \\ -1/6 & -1/3 & 1 \end{bmatrix}$, $x = \begin{bmatrix} x_1 \\ x_2 \\ x_3 \end{bmatrix}$

16. $x = c_1 \begin{bmatrix} 3 \\ -2 \end{bmatrix} e^{-t} + c_2 \begin{bmatrix} 1 \\ 0 \end{bmatrix} e^{t}$.

18. $x = c_1 \begin{bmatrix} 1 \\ -1 \\ 0 \end{bmatrix} e^{t} + c_2 \begin{bmatrix} 1 \\ 0 \\ -2 \end{bmatrix} e^{t} + c_2 \begin{bmatrix} 2 \\ 2 \\ 1 \end{bmatrix} e^{10t}$

20. $x = c_1 \begin{bmatrix} 1 \\ 0 \\ 0 \end{bmatrix} + 2\text{Re}\left[(c_2 + ic_3) \begin{bmatrix} \frac{1}{2}(1 + i) \\ 1 \\ -1 \end{bmatrix} e^{-it} \right]$

   or

41

20. Cont'd...

$$x_1 = c_1 e^t + (c_2 - c_3) \cos t + (c_2 + c_3) \sin t$$

$$x_2 = 2 c_2 \cos t + 2 c_3 \sin t$$

$$x_3 = 2 c_3 \cos t - 2 c_2 \sin t \ .$$

22. $x = c_1 \begin{bmatrix} 0 \\ 1 \\ -1 \end{bmatrix} e^{-2t} + c_2 \begin{bmatrix} 1 \\ 0 \\ 0 \end{bmatrix} e^{4t} + c_3 \begin{bmatrix} 0 \\ 1 \\ 1 \end{bmatrix} e^{4t} \ .$

30. $x = c_1 \begin{bmatrix} 1 \\ -1 \end{bmatrix} e^{-t} + c_2 \begin{bmatrix} 1 \\ 0 \end{bmatrix} e^{-t} + c_2 \begin{bmatrix} 1 \\ -1 \end{bmatrix} te^{-t} \ .$

32. $x_1 = c_1 e^t + c_2 t e^t + c_3 t^2 e^t$

$$x_2 = c_2 e^t - \frac{4}{3} c_3 e^t + 2 c_2 t e^t$$

$$x_3 = \frac{2}{3} c_3 e^t \ .$$

38. Expand e^{tA} using the fact that

$$A = \begin{bmatrix} 0 & -k \\ k & 0 \end{bmatrix}, \quad A^2 = \begin{bmatrix} -k^2 & 0 \\ 0 & -k^2 \end{bmatrix}, \quad A^3 = \begin{bmatrix} 0 & k^3 \\ -k^3 & 0 \end{bmatrix},$$

$$A^4 = \begin{bmatrix} k^4 & 0 \\ 0 & k^4 \end{bmatrix},$$

and $A^n = A^{n+4}$, $n = 1, 2, \dots,$. Sum the powers of t in each of the four elements of e^{tA} and identify them with the series for sin kt and coskt. The result for e^{tB} follows in similar fashion.

40. $dM/dt = Ae^{tA}e^{-tA} - e^{tA}Ae^{-tA}$, but by Problem 37 this is equivalent to $dM/dt = Ae^{tA}e^{-tA} - Ae^{tA}e^{-tA} = 0$. Thus $M(t) = $ constant vector $= M(0) = e^0 e^0 = I$. Hence $e^{tA}e^{-tA} = I$, so that e^{-tA} is the inverse of e^{tA} and thus e^{tA} is non singular with $(e^{tA})^{-1} = e^{-tA}$.

42. $\int e^{-tA} dt = \int \left[I - \frac{t}{1!} A + \frac{t^2}{2!} A^2 - \frac{t^3}{3!} A^3 + ... \right] dt$

$= tI - \frac{t^2}{2!} A + \frac{t^3}{3!} A^2 - \frac{t^4}{4!} A^3 + ... + C$

$= A^{-1} - A^{-1} e^{-tA} + C$

$= -A^{-1} e^{-tA}$, to within an arbitrary constant matrix.

A^{-1} and A^m commute, so $-A^{-1} e^{-tA} = -e^{-tA} A^{-1}$.

44. The result is an immediate consequence of induction on m, starting with m = 1 for which the result is certainly true.

46. The result is immediate, because each element x_i of x is a linear combination of the elements of the ith row of $e^{(t - t_0)A}$ and the elements of x_0. Thus x_i is a linear combination of an element from each column of $e^{(t - t_0)A}$.

48. As B and C commute, $e^{t(A + B)} = e^{tA} e^{tB}$, but from Problem 38

$$e^{tB} = \begin{bmatrix} e^{at} & 0 \\ 0 & e^{at} \end{bmatrix} \quad \text{and} \quad e^{tC} = \begin{bmatrix} \cos kt & -\sin kt \\ \sin kt & \cos kt \end{bmatrix}$$

The solution then follows from $x = e^{t(B + C)} x_0 = e^{tB} e^{tC} x_0$.

50. $A = P\Lambda P^{-1}$, $A^2 = P\Lambda P^{-1} P\Lambda P^{-1} = P\Lambda^2 P^{-1}$ etc.

$e^{tA} = I + \frac{t}{1!} A + \frac{t^2}{2!} A^2 + ...$

$= I + \frac{t}{1!} P\Lambda P^{-1} + \frac{t^2}{2!} P\Lambda^2 P^{-1} + ...$

$= P e^{t\Lambda} P^{-1}$.

$$\Lambda^m = \begin{bmatrix} \lambda_1^m & & & & 0 \\ & \lambda_2^m & & & \\ & & \cdot & & \\ & & & \cdot & \\ & & & & \cdot \\ 0 & & & & \lambda_n^m \end{bmatrix}$$, so by summing terms we find

50. Cont'd...

$$e^{t\Lambda} = I + \frac{t}{1!} \Lambda + \frac{t^2}{2!} \Lambda^2 + \dots + \quad = \begin{bmatrix} e^{\lambda_1 t} & & & & 0 \\ & e^{\lambda_2 t} & & & \\ & & \cdot & & \\ & & & \cdot & \\ 0 & & & & \cdot \\ & & & & e^{\lambda_n t} \end{bmatrix} .$$

The last result follows by substituting for $e^{t\Lambda}$ in

$$e^{tA} = P e^{t\Lambda} P^{-1} .$$

Section 6.2

2. $\lambda_1 = 7$, $\lambda_2 = -4$, $\quad P = \begin{bmatrix} 1 & -6 \\ 1 & 5 \end{bmatrix}$, $\quad P^{-1} = \frac{1}{11} \begin{bmatrix} 5 & 6 \\ -1 & 1 \end{bmatrix}$,

$\Lambda = \begin{bmatrix} 7 & 0 \\ 0 & -4 \end{bmatrix}$; $x_1 = c_1 e^{7t} - 6c_2 e^{-4t} + \frac{11}{14}$,

$x_2 = c_1 e^{7t} + 5c_2 e^{-4t} - \frac{55}{28}$.

4. $\lambda_1 = 1$, $\lambda_2 = -1$, $\quad P = \begin{bmatrix} 1 & 1 \\ -1 & 2 \end{bmatrix}$, $\quad P^{-1} = \frac{1}{3} \begin{bmatrix} 2 & -1 \\ 1 & 1 \end{bmatrix}$,

$\Lambda = \begin{bmatrix} 1 & 0 \\ 0 & -1 \end{bmatrix}$; $\quad x_1 = c_1 e^t + c_2 e^{-t} - t + 3$,

$x_2 = -c_1 e^t + 2c_2 e^{-t} + 4t + 3$; $c_1 = 1$, $c_2 = -1$.

6. $\lambda_1 = 1$, $\lambda_2 = -1$, $\lambda_3 = -2$, $\quad P = \begin{bmatrix} 1 & 5 & 2 \\ 0 & 2 & 0 \\ 0 & -2 & -3 \end{bmatrix}$

$P^{-1} = \frac{1}{6} \begin{bmatrix} 6 & -11 & 4 \\ 0 & 3 & 0 \\ 0 & -2 & -2 \end{bmatrix}$, $\quad \Lambda = \begin{bmatrix} 1 & 0 & 0 \\ 0 & -1 & 0 \\ 0 & 0 & -2 \end{bmatrix}$;

$x_1 = c_1 e^t - 2c_2 e^{-2t} + 5c_3 e^{-t} + 3t + \frac{1}{2}$,

$x_2 = 2c_3 e^{-t} + t$, $x_3 = 3c_2 e^{-2t} - 2c_3 e^{-t} - \frac{1}{2}t + \frac{5}{4}$;

$c_1 = -\frac{4}{3}$, $c_2 = -\frac{5}{12}$, $c_3 = 0$.

8. $\lambda_1 = -1$, $\lambda_2 = -3$, $\lambda_3 = -4$, $\quad P = \begin{bmatrix} 1 & 1 & 0 \\ 0 & -2 & 1 \\ 0 & 1 & -1 \end{bmatrix}$,

$P^{-1} = \begin{bmatrix} 1 & 1 & 1 \\ 0 & -1 & -1 \\ 0 & -1 & -2 \end{bmatrix}$ $\Lambda = \begin{bmatrix} -1 & 0 & 0 \\ 0 & -3 & 0 \\ 0 & 0 & -4 \end{bmatrix}$; $x_1 = 3e^{-t} - e^{-3t} + 7/3$,

8. Cont'd...

$x_2 = 2e^{-3t} - e^{-4t} + 7/12$, $x_3 = -e^{-3t} + e^{-4t} + 1/12$.

10. $\lambda_1 = 2$, $\lambda_2 = -3$, $\lambda_3 = -3$, $Q_2 = \begin{bmatrix} 1 & 0 & 0 \\ 0 & 1 & 0 \\ 0 & -3 & 1 \end{bmatrix}$, $Q_2^{-1} = \begin{bmatrix} 1 & 0 & 0 \\ 0 & 1 & 0 \\ 0 & 3 & 1 \end{bmatrix}$

$T_2 = \begin{bmatrix} 2 & 0 & 0 \\ 0 & -3 & 1 \\ 0 & 0 & -3 \end{bmatrix}$; $x_1 = c_1 e^{2t}$, $x_2 = c_2 e^{-3t} + c_3 t e^{-3t} + \frac{2}{25} \cos t +$

$\frac{3}{50} \sin t$, $x_3 = (c_3 - 3c_2) e^{-3t} - 3c_3 t e^{-3t} + \frac{3}{50} \cos t - \frac{27}{25} \sin t$.

12. If $w = x_1 - x_2$, then $\dot{w} = Aw$. Now $w(t_o) = x_1(t_o) - x_2(t_o) \equiv 0$. Thus as the general solution $w = c_1 w_1 + c_2 w_2 + ... + c_n w_n$, with w_1, w_2 , ..., w_n a basis for the solution space, it follows that $c_1 = c_2 = ... = c_n = 0$. Thus $w \equiv 0$ for $t \geq t_o$, and so $x_1 \equiv x_2$ and hence the solution is unique.

14. $\int_{t_o}^{t} \frac{d}{d\tau} (e^{-\tau A} x) d\tau = \int_{t_o}^{t} e^{-\tau A} f(\tau) d\tau$ and so

$e^{-tA} x(t) = e^{-t_o A} x(t_o) + \int_{t_o}^{t} e^{-tA} f(\tau) d\tau$.

The result now follows after pre–multiplication by e^{tA}, using the fact that when A and B commute, $e^{\alpha A} e^{\beta B} = e^{\alpha A + \beta B}$ (Prob. 41, Sec. 6.1), and taking e^{tA} under the integral sign.

16. $\lambda_1 = -1$, $\lambda_2 = -2$, $\lambda_3 = -3$, $e^{tA} = Pe^{tA}P^{-1}$ with $P = \begin{bmatrix} 1 & 1 & 1 \\ -1 & 0 & -1 \\ 0 & 1 & 2 \end{bmatrix}$,

$P^{-1} = \begin{bmatrix} \frac{1}{2} & -\frac{1}{2} & -\frac{1}{2} \\ 1 & 1 & 0 \\ -\frac{1}{2} & -\frac{1}{2} & \frac{1}{2} \end{bmatrix}$

$\Lambda = \begin{bmatrix} -1 & 0 & 0 \\ 0 & -2 & 0 \\ 0 & 0 & -3 \end{bmatrix}$, $x_1 = -e^{-2t}$, $x_2 \equiv 0$, $x_3 = 1 - e^{-2t}$.

18. (i) $A = UAU^{-1}$, so $U^{-1}A = U^{-1}UAU^{-1} = AU^{-1}$.

(ii) $dV/dt = Ae^{tA}Ue^{-tA} - e^{tA}UAe^{-tA}$, but A, e^{tA} and e^{-tA} commute, so $dV/dt = e^{tA}(AU - UA)e^{-tA} = 0$ by (i), for all t. Thus $V(t) = V(0) \equiv$ const; but $V(0) = U$ and thus $V \equiv U$ for all t. The penultimate result follows by using integration by parts after noticing that

$$\int_0^t e^{-\tau A} e^{p\tau} C \, d\tau = \int_0^t e^{-\tau A} \frac{d\left[\frac{1}{p} e^{p\tau}\right]}{d\tau} \, d\tau .$$

The final result follows by grouping terms and then by applying (i) and (iii) to $e^{tA}(pI - A)^{-1}e^{Pt}e^{-At}C$.

Section 6.3

2. $x(t) = c_1 \begin{bmatrix} 2 \\ 2 \\ -1 \end{bmatrix} \cosh(\sqrt{3}t + \alpha_1) + c_2 \begin{bmatrix} -1 \\ 2 \\ 2 \end{bmatrix} \cosh(\sqrt{6}t + \alpha_2)$

$+ c_3 \begin{bmatrix} 2 \\ -1 \\ 2 \end{bmatrix} \cosh(3t + \alpha_3)$.

4. Natural angular frequencies $\omega_1 = 2\sqrt{2/3}$, $\omega_2 = 2\sqrt{2}$. Normal modes :

$x^{(1)}(t) = \begin{bmatrix} 1 \\ 3 \end{bmatrix} \cos\left[2\sqrt{\frac{2}{3}}\, t + \alpha_1\right]$, $x^{(2)}(t) = \begin{bmatrix} 1 \\ -1 \end{bmatrix} \cos(2\sqrt{2}t + \alpha_2)$

$x_1(t) = \frac{1}{2}\left[\cos 2\sqrt{\frac{2}{3}}\, t + \cos 2\sqrt{2}\, t\right]$

$x_2(t) = \frac{1}{2}\left[3 \cos 2\sqrt{\frac{2}{3}}\, t - \cos 2\sqrt{2}\, t\right]$.

6. Natural angular frequencies $\omega_1 = \frac{1}{2}\left[5 - \sqrt{17}\right]^{1/2} p$, $\omega_2 = \frac{1}{2}\left[5 + \sqrt{17}\right]^{1/2} p$. Natural frequencies of oscillation $f_1 = \frac{1}{4\pi}\left[5 - \sqrt{17}\right]^{1/2} p$,

$f_2 = \frac{1}{4\pi}\left[5 + \sqrt{17}\right]^{1/2} p$.

8. Normal modes $x^{(1)}(t) = \begin{bmatrix} 1 \\ -\sqrt{2} \end{bmatrix} \cos(\omega_1 t + \alpha_1)$, $x^{(2)}(t) = \begin{bmatrix} 1 \\ \sqrt{2} \end{bmatrix} \cos(\omega_2 t + \alpha_2)$,

where $\omega_1 = (a - \sqrt{2}b)^{1/2}p$, $\omega_2 = (a + \sqrt{2}b)^{1/2}p$.

First mode is SHM along $x + \sqrt{2}y = 0$ with angular frequency ω_1.

Second mode is SHM along $x - \sqrt{2}y = 0$ with angular frequency ω_2.

10. Natural angular frequencies $\omega_1 = \omega_0$, $\omega_2 = \left[\omega_0^2 + \dfrac{2k}{m}\right]^{1/2}$.

First mode 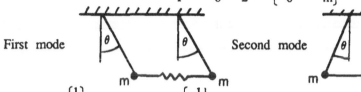 Second mode

$x(t) = \frac{1}{2} a \begin{bmatrix} 1 \\ 1 \end{bmatrix} \cos \omega_0 t + \frac{1}{2} a \begin{bmatrix} 1 \\ -1 \end{bmatrix} \cos \omega_2 t$.

12. $q_1(t) = E_0 \left[\frac{3\sqrt{6}}{8} \sin \sqrt{\frac{2}{3}} t + \frac{\sqrt{2}}{8} \sin \sqrt{2}\, t - \sin t \right]$

$q_2(t) = E_0 \left[\frac{9\sqrt{6}}{8} \sin \sqrt{\frac{2}{3}} t - \frac{\sqrt{2}}{8} \sin \sqrt{2}\, t - 2 \sin t \right]$.

16. $y = x_2 = -2c_1 e^{-2t} - c_2 e^{-t} - c_3 t e^{-t}$, and corresponding to this,

$x_1 = x = c_1 e^{-2t} + c_2 e^{-t} + c_3(1 + t)e^{-t}$,

$x_3 = \dot{x}_2 = 4c_1 e^{-2t} + c_2 e^{-t} + c_3(t - 1)e^{-t}$.

The natural oscillatory behaviour has been suppressed and changed to one of purely exponential decay.

18. The analysis proceeds as in Example 4, but now the conditions at the ends of the system become $x_0 = x_N = 0$. As a result $\theta_s = s\pi/N$, from which the required results then follow.

20. $\exp\{\frac{1}{2}tP\} = \begin{bmatrix} 1 & \frac{1}{2}t & \frac{1}{2}t\left[1 + \frac{t}{4}\right] \\ 0 & 1 & \frac{1}{2}t \\ 0 & 0 & 1 \end{bmatrix}$,

20. Cont'd...

$$\exp\{-\tfrac{1}{2}tP\} = \begin{bmatrix} 1 & -\tfrac{1}{2}t & \tfrac{1}{2}\left[\tfrac{t}{4}-1\right] \\ 0 & 1 & -\tfrac{1}{2}t \\ 0 & 0 & 1 \end{bmatrix}$$

General solution:

$$x_1 = c_1 \cos(t + \alpha_1) - \frac{c_2 t}{2} \cos(t + \alpha_2)$$
$$+ \frac{c_3 t}{2}\left[\tfrac{t}{4}-1\right]\cos(t+\alpha_3) + \tfrac{t}{2}$$
$$x_2 = c_2 \cos(t+\alpha_2) - \frac{c_3 t}{2}\cos(t+\alpha_3) + 1$$
$$x_3 = c_3 \cos(t+\alpha_3) .$$

Solution of initial value problem:

$$x_1 = \tfrac{t}{2}\left[\cos t + 1\right], \quad x_2 = 1 - \cos t, \quad x_3 \equiv 0.$$

Section 6.4

2. (a) $(0, 0)$ (b) $(2, 0)$ 4. (a) $(1/3, 0)$ (b) $(\tfrac{1}{2}, 0)$.

6. $(0, 0)$ if b, c have the same sign; $(0, 0)$, $(\pm \sqrt{b/c}, 0)$ if b, c have
   opposite signs.

8. Trajectories have a common
   tangent at the origin.

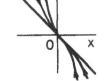

10. Spirals directed away from
    the origin.

12. $k = [(1 - x^2)y - x]/y$, with $k = dy/dx$.

16. Equation of trajectories is $\tfrac{1}{2}X^2 = \cos\theta + c$; Equations of separatrices are
    $\tfrac{1}{2}X^2 = \cos\theta \pm 1$.

20. Same pattern as Fig. 6.30, but with center at (−3, 0) and saddle point at (−1, 0).

22. Lightly damped pendulum with an oscillatory solution. Equilibrium points on the x−axis at x = ±nπ, n = 0, 1, 2, ... Points at even multiples of π are stable spirals and points at odd multiples of π are saddle points.

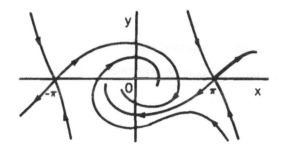

(y = dx/dt)

24.

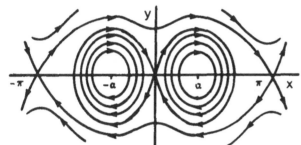

Extend using symmetry about the y−axis and periodicity with period 2π.

26. Equilibrium in second cycle at x = 0, y = 0.

28. Equilibrium in fourth cycle at x = 0, y = 0.

30. Equilibrium in second cycle at x = 1, y = 0 .

32. (a) No, (b) Yes, (c) No.   34. (a) No, (b) No, (c) Yes.

36. The cubic will have roots with negative real parts if

$$a > 0 , \quad \begin{vmatrix} a & 1 \\ b & 0 \end{vmatrix} > 0 , \quad \begin{vmatrix} a & 1 & 0 \\ b & 0 & a \\ 0 & 0 & b \end{vmatrix} > 0, \text{ or}$$

36   Cont'd...

a > 0, b < 0 which are possible and $-b^2 > 0$, which is impossible. Thus not all of the roots can have negative real parts. Consequently the general solution of the differential equations cannot be asymptotically stable.

38.   No, because P(x) has two positive zeros.

40.   No, because q(x) < 0.   42.   Center

44.   Unstable improper node with an exceptional direction.

46.   Center   48.   Saddle point   50.   Saddle point

52.   Equilibrium points at (0, 0) and $\left[\dfrac{b}{c}, \dfrac{a}{c}\right]$. The saddle point at the origin is of no interest as there are neither prey nor predators. The point $\left[\dfrac{b}{c}, \dfrac{a}{c}\right]$ is a center. Cyclic behaviour of the predator and prey population occurs around this point. More foxes mean less rabbits, which reduces the fox population, after which the rabbit population will increase, and so on.

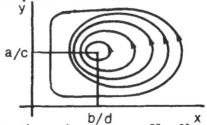

56.   Stable attractive node.   58.   Unstable proper node

60.   Unstable degenerate node without an exceptional direction.

62.   Consider the annular region $2 \le r \le 4$; $\dfrac{dr}{dt} = r(9 - r^2)$, $\dfrac{d\theta}{dt} = 9$;
      $r = 3(1 + ae^{-18t})^{-1/2}$, $\theta = 9t + t_0$; limit cycle is circle r = 3.

64.   Consider the annular region $\frac{1}{2} \le r \le 2$; $\dfrac{dr}{dt} = r(1 - r^2)$; $\dfrac{d\theta}{dt} = -4$
      $r = (1 + ae^{-2t})^{-1/2}$, $\theta = -4t + t_0$; limit cycle is circle r = 1.

66.   This represents the limit of the flow in Prob. 65 as a $\rightarrow$ 0, bringing the two nodes together at the origin. The confluence of these two nodes is called a **dipole**.

Section 6.5

2.

n	x_n	k_{1n}	K_{1n}	k_{2n}	K_{2n}
0	0	0.445885	−0.099499	0.476652	−0.087640
1	0.2	0.518270	−0.060395	0.561516	−0.019917
2	0.4	0.628394	0.048795	0.698139	0.130413
3	0.6	0.810290	0.252733	0.924683	0.374358
4	0.8	1.094406	0.520961	1.255223	0.618199

n	x_n	k_{3n}	K_{3n}	k_{4n}	K_{4n}
0	0	0.481388	−0.085964	0.518143	−0.060570
1	0.2	0.571600	−0.013153	0.628861	0.049192
2	0.4	0.716553	0.145312	0.812055	0.254286
3	0.6	0.951491	0.396255	1.098190	0.523731
4	0.8	1.279544	0.626653	1.452488	0.655002

n	x_n	k_n	K_n	y_n	z_n
0	0	0.480018	−0.084546	1.5	0.5
1	0.2	0.568894	−0.012890	1.980018	0.415454
2	0.4	0.711639	0.142422	2.548911	0.402564
3	0.6	0.943471	0.386281	3.260550	0.544985
4	0.8	1.269405	0.610945	4.204021	0.931267

4.

n	x_n	k_{1n}	K_{1n}	k_{2n}	K_{2n}
0	0.5	0.200000	0.250000	0.222500	0.252533
1	0.6	0.247661	0.257283	0.272908	0.262454
2	0.7	0.301383	0.270350	0.329969	0.279152
3	0.8	0.362518	0.291363	0.395212	0.305097
4	0.9	0.432826	0.323293	0.470632	0.343838

n	x_n	k_{3n}	K_{3n}	k_{4n}	K_{4n}
0	0.5	0.223752	0.253532	0.247728	0.257293
1	0.6	0.274429	0.263476	0.301451	0.270378
2	0.7	0.331839	0.280232	0.362590	0.291411
3	0.8	0.397533	0.306330	0.432904	0.323361
4	0.9	0.473550	0.345424	0.514723	0.370647

4. Cont'd...

n	x_n	k_n	K_n	y_n	z_n
0	0.5	0.223372	0.253237	2.0	0.0
1	0.6	0.273964	0.263254	2.223372	0.253237
2	0.7	0.331265	0.280088	2.497336	0.516490
3	0.8	0.396819	0.306263	2.828601	0.796578
4	0.9	0.472652	0.345410	3.225420	1.102842

6.

n	x_n	k_{1n}	K_{1n}	k_{2n}	K_{2n}
0	−0.5	0.031706	0.000000	0.031656	0.036829
1	−0.3	0.035309	0.066530	0.038890	0.096346
2	−0.1	0.045890	0.119692	0.053090	0.143133
3	0.1	0.063363	0.160650	0.073938	0.178249
4	0.3	0.086681	0.190440	0.099459	0.202728

n	x_n	k_{3n}	K_{3n}	k_{4n}	K_{4n}
0	−0.5	0.033674	0.033152	0.035153	0.066635
1	−0.3	0.040648	0.093006	0.045722	0.119799
2	−0.1	0.054573	0.140069	0.063177	0.160763
3	0.1	0.075003	0.175431	0.086481	0.190563
4	0.3	0.099929	0.200221	0.112557	0.210410

n	x_n	k_n	K_n	y_n	z_n
0	−0.5	0.032920	0.034433	0.0	−1.0
1	−0.3	0.040018	0.094172	0.032920	−0.965567
2	−0.1	0.054066	0.141143	0.072937	−0.871395
3	0.1	0.074621	0.176429	0.127003	−0.730252
4	0.3	0.099669	0.201125	0.201624	−0.553823

Chapter 7.   Laplace Transform and z–transform.

Section 7.1

2. $(3 + 4s)/(s^2 + 3s)$

4. $(s^2 - k^2 + i2sk)/(s^2 + k^2)^2$

6. $(1 - e^{-s})^2/s$

8. $(s - 1)/(s^2 - 2s - 3)$

10. $e^{-as}/(s^2 + 1)$

12. $(s + e^{-s} - 1)/s^2$

14. $(a + (b - a)e^{-\alpha s} - be^{-\beta s})/s$

16. $2abs/\{[s^2 + (a+b)^2][s^2 + (a-b)^2]\}$

18. $2ks^2/(s^2 + k^2)^2$

20. $k^3/\{s^2(s^2 + k^2)\}$

22. $(s - 3)/(s^2 - 2s + 10)$

24.   Use $\sin^2 kt = \frac{1}{2}(1 - \cos 2kt)$ to obtain $2k^2/\{s(s^2 + 4k^2)\}$

26. $2 \sin 3t$

28. $3t \cos 2t$

30. $8 \sinh 2t$

32. $\frac{1}{3} e^{-t} \cos t$

34. $3t + \frac{1}{6} t \sin kt$

36. $\frac{t^2}{2}(e^{2t} + e^{-st}) = t^2 \cosh 2t$

Section 7.2

2. $\dfrac{4(s - 2)}{((s - 2)^2 + 4)^2}$

4. $\dfrac{6}{(s + 2)^2}$

6. $\dfrac{s - 4}{(s - 4)^2 - 9}$

8. $\frac{1}{2} e^{-2t} \sin 2t$

10. $e^{2t} \sinh 2t = \frac{1}{2}(e^{4t} - 1)$

12. $e^{-t}(2t^2 + 3 \cos 4t)$

14. $(2 + e^{2t}) \cos 3t$

16.

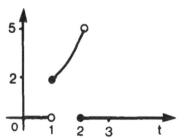

18.

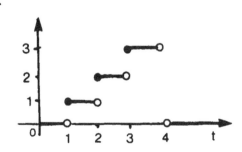

20.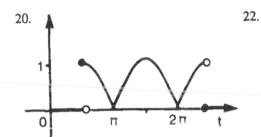

22.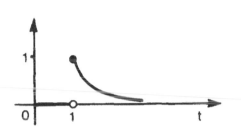

24.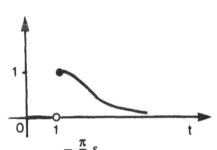

26. $F(s) = \dfrac{se^{-\frac{\pi}{2}s}}{s^2 + 4}$

28. $F(s) = \dfrac{e^{-\pi s}}{s^2 + 4}$

30. $\mathscr{L}\{u(t-2)(t+1)\} = \mathscr{L}\{u(t-2)(t-2)\} = 3\,\mathscr{L}\{u(t-2)\} = \dfrac{e^{-2s}}{s^2} + \dfrac{3e^{-s}}{s}$.

32. $f(t) = \frac{1}{4}u(t-1)\{(t-1)\sin 4(t-1)\}$

34. $f(t) = 3u(t-4)\cosh 2(t-4)$

36. $f(t) = u(t-2)\sinh 2(t-2) + 2u(t-4)\cosh(t-4)$

38. $f(t) = u(t-1)\{2(t-1)\} + u(t-2)\{(t-2)\,e^{-3(t-2)}\}$

40. $\mathscr{L}\{e^t \cos \frac{3}{2} t\} = \dfrac{4s-4}{4s^2 - 8s + 13}$

42. $\mathscr{L}\{f(\frac{3}{2} t)\} = \dfrac{6 + 4se^{-\frac{2s}{3}}}{4s^2 + 24s - 27}$

44. $\mathscr{L}\{t \sin 2t \sinh 2t\} = \dfrac{32s}{(s^2 + 4)^2(s^2 - 4)}$

48. $\mathscr{L}\{t^3 e^t\} = \dfrac{6}{(s-1)^4}$, for $s > 1$

50. $\mathcal{L}\{\cosh^2 t\} = \dfrac{2 - s^2}{s(4 - s^2)}$, for s > 2

52. $\mathcal{L}\{t \cos 3t\} = \dfrac{s^2 - 9}{(s^2 + 9)^2}$, for s > 0

54. $\mathcal{L}\{f\,'\} = \dfrac{2s^2}{(s^2 + 1)^2}$, $\mathcal{L}\{f''\} = \dfrac{2s^3}{(s^2 + 1)^2}$

56. $\mathcal{L}\{f\,'\} = \dfrac{2s}{s^2 - 2s + 5}$, $\mathcal{L}\{f''\} = \dfrac{4s - 10}{s^2 - 2s + 5}$

58 $\mathcal{L}\{f\,'\} = \dfrac{4s}{s^2 - 16}$, $\mathcal{L}\{f''\} = \dfrac{16}{s^2 - 16}$

60. $y(t) = 4e^{2t} - e^t$

62. $y(t) = \dfrac{1}{10}(\sin t + 2 \cos t) - \dfrac{1}{20} e^{-t}(4 \cos 2t - 3 \sin 2t)$

64. $y(t) = (1 - 2t + 2t^2)e^{2t}$

66. $y(t) = 19 - 17t - 18e^{-t} + 8t^2 - te^{-t}$

68. Shift time origin by setting $\tau = t - 2$ to obtain

$$u\,'' + 4u\,' + 3u = 4e^{-\tau} ,\text{ with } u(0) = 0,\ u\,'(0) = 2,$$

where $x(t) = x(\tau + 2) = u(\tau)$. This has the solution

$$u(\tau) = 2e^{\tau} + e^{-2\tau}(6 \sin\tau - 2 \cos \tau) \text{ for } \tau > 0.$$

Required solution follows by using $\tau = t - 2$, for $\tau > 0$.

70. $x(t) = \cos 2t + \dfrac{1}{4}u(t - 1)\{1 - \cos 2(t - 1)\}$

72. $x(t) = -1 + \cos t + u(t - 1)[1 - \cos(t - 1)]$.

74. $x(t) = e^{-t} - e^{-2t} + \dfrac{1}{2}u(t - 2)[1 - 2e^{-(t - 2)} + e^{-2(t - 2)}]$.

76. $x(t) = \dfrac{1}{2} - \dfrac{1}{2} e^{-2t} + u(t - 1)[\dfrac{1}{4} + \dfrac{1}{2}(t - 1) + e^{-(t - 1)} - \dfrac{1}{4} e^{-2(t - 1)}]$.

78. $x(t) = e^{-2t} + 4te^t - e^t$, $y(t) = \frac{2}{3} e^{-2t} + 2te^t - \frac{2}{3} e^t$.

80. $x(t) = (t - 1)e^{-2t}$, $y(t) = (2 - t) e^{-2t}$.

82. $x(t) = \frac{1}{3} (e^{-t} + 2e^{2t})$, $y(t) = \frac{1}{3}(2e^{2t} - 5e^{-t})$, $z(t) = \frac{2}{3} (2e^{-t} + e^{2t})$.

84. $x(t) = \frac{1}{2} (\sinh t - \sin t)$, $y(t) = \frac{1}{2} (\sinh t + \sin t)$

86. $\mathscr{L} \{t^{3/2} e^{-t}\} = \dfrac{3\sqrt{\pi}}{4(s - 1)^{5/2}}$.

88. $\mathscr{L} \{t^n e^{at}\} = \dfrac{n!}{(s - a)^{n + 1}}$.

92. $\dfrac{n!}{s(s - a)^{n + 1}}$

94. $\dfrac{2k^3}{s(s^4 - k^4)}$

96. $\dfrac{k^3}{s^3(s^2 + k^2)}$

98. $\dfrac{1}{6} t^3 e^t$

100. $\dfrac{2}{k^3} (\sinh kt - \sin kt)$

102. $t \cos kt$

104. $\mathscr{L} \{ \dfrac{\sin t}{t} \} = \dfrac{\pi}{2} - \arctan s$.

106. $\mathscr{L} \{\cos kt\} = \dfrac{s}{s^2 + k^2}$

108. $\sinh kt$

110. $\cosh kt$

112. $\dfrac{1}{s^2} \tanh ks$

114. $\dfrac{1}{(s^2 + 1) (1 - e^{-\pi s})}$

116. $\dfrac{1}{2s} \dfrac{e^{-\frac{ks}{2}}}{\cosh(\frac{ks}{2})}$

122. $1 - \cos t$

126. $\dfrac{1}{(s^2 + 1)^2}$

128. $\dfrac{8s^2}{(s^2 + 1)^2(s^2 - 16)}$

130. $\dfrac{1}{k^3} (kt - \sin kt)$

132. $u (t - k) (t - k)$

134. Result (i) follows from the linearity of the definite integral. Result (ii) is most easily demonstrated by example. Taking $f(t) = e^t$, it follows that $f*1 = e^t - 1 \neq f(t)$.

136. $f(0) = f\,'(0) = 0;\ f(t) = \dfrac{1}{k^2}\,(1 - \cos kt).$

138. $f(0) = f\,'(0) = 0;\ f(t) = \tfrac{1}{2}\,u\,(t - 3)\ e^{-(t - 3)}\sin 2(t - 3).$

140. $f(0) = 0,\ f\,'(0) = 1;\ f(t) = t\cos t.$

142. $\displaystyle\int_{-\infty}^{\infty} \Delta_k(t)\ dt = 1$ and $\displaystyle\lim_{k\to\infty}\ \Delta_k(t) = 0$ for $t \neq 0.$

144. $I = \displaystyle\int_{-\infty}^{\infty} \dfrac{\sin^2 t}{1 + 2t^2}\ \delta(t + \pi)\ dt + \int_{-\infty}^{\infty}\dfrac{\sin^2 t}{1 + 2t^2}\ \delta(t - 4\pi)\ dt = 0 + 0 = 0$

146. $I = \displaystyle\int_{1}^{\infty} \dfrac{dt}{1 + t^2} = \pi/4,$ because $u\,(t - 1)\ \delta(t + 1) \equiv 0$ for all $t.$

148. Let $g(t)$ be an arbitrary continuous function. Then

$$\int_{-\infty}^{\infty} g(t)\ \{f(t)\ \delta(t - t_0) - f(t_0)\ \delta(t - t_0)\}\ dt = g(t_0)\ f(t_0) - g(t_0)\ f(t_0)$$

$= 0,$ from which the required result follows by use of (28). The second result follows by setting $t_0 = 0$ and $f(t) = t.$

150. $\delta((t - t_1)\,(t - t_2))$ has singularities at t_1 and t_2 and is zero elsewhere. At t_1 it is the delta function $\delta((t_1 - t_2)\tau)$ and at t_2 the delta function $\delta((t_2 - t_1)\tau).$ Adding these expressions and using the result of Ex. 18 gives the required result.

152. The result is true for $n = 1.$ Assuming it is true for $n,$ differentiation with respect to t of $t^n\ \delta^{(n)}(t) = (-1)^n n!\ \delta(t)$ followed by multiplication by t and use of the results for n and 1 gives

$$t^{n + 1}\delta^{(n + 1)}(t) = (-1)^{n + 1}(n + 1)!\ \delta(t).$$

This is the result for $n + 1,$ so the general result follows by induction. The sifting property is

$$\int_{-\infty}^{\infty} f(t)\ t^n\ \delta^{(n)}(t)\ dt = (-1)^n n!\ f(0).$$

154. $\dfrac{d}{dt}\,|t| = \dfrac{d}{dt}\,\{t[2u\,(t) - 1]\} = 2u\,(t) - 1 + 2t\,\dfrac{d}{dt}\,u\,(t)$

$= 2u\,(t) - 1 + 2t\delta(t) = 2u\,(t) - 1,$ because $t\delta(t) = 0.$

154. Cont'd...

Thus $\frac{d^2}{dt^2}|t| = \frac{d}{dt}[2\,\mathcal{u}(t) - 1] = 2\delta(t)$. The second result follows in similar fashion.

156. (i) $\frac{s^2}{s^2 - k^2} = 1 + \frac{k^2}{s^2 - k^2}$, so $\mathcal{L}^{-1}\left\{\frac{s^2}{s^2 - k^2}\right\} = \delta(t) + k \sinh kt$.

As $\mathcal{L}^{-1}\left\{\frac{1}{s}\right\} = 1$, $\mathcal{L}^{-1}\left\{\frac{s}{s^2 - k^2}\right\} = (\delta(t) + k \sinh kt)*1 = \cosh kt$.

158. $\mathcal{L}^{-1}\left\{3s + 1 + \frac{2}{s - 1} + \frac{1}{s + 2}\right\} = 3\delta'(t) + \delta(t) + 2e^t + e^{-2t}$.

160. $f(t) = \mathcal{u}(t) + 2\sum_{n=1}^{\infty}(-1)^n\,\mathcal{u}(t - 2nk)$, for $t \geq 0$.

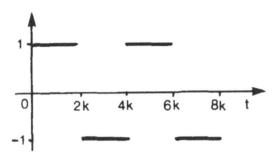

$\mathcal{L}\{f(t)\} = \frac{1}{s} + 2\sum_{n=1}^{\infty}(-1)^n\frac{e^{-2nks}}{s} = \frac{1}{s} - \frac{2e^{-2ks}}{1 + e^{-2ks}} = \frac{1}{s}\tanh ks$ (an infinite geometric series with common ratio $-e^{-2ks}$).

162. $f(t) = 2\sum_{n=0}^{\infty}(-1)^n\,\mathcal{u}(t - (2n + 1)k)$, for $t \geq 0$.

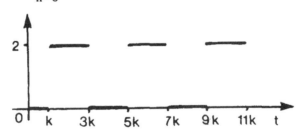

$\mathcal{L}\{f(t)\} = \frac{2}{s}\sum_{n=0}^{\infty}(-1)^n\,e^{-2nks} = \frac{1}{s\cosh ks}$ (an infinite geomatric series with common ratio e^{-2ks}).

164. $x(t) = \frac{1}{4} + \frac{1}{4} e^{-t}(\sqrt{3} \sin \sqrt{3} t - \cos \sqrt{3} t)$.

166. $x(t) = \frac{1}{8} (5 \sin 2t - 2t \cos 2t)$.

168. $x(t) = \frac{1}{2} (3e^{-t} - e^{t})$, $y(t) = \frac{1}{2} (3e^{-t} + e^{t} - 2)$.

170. $y(t) = 3(\sin t - t \cos t)$.

172. $y(t) = \frac{1}{40} e^{5t}(\sin 4t - 4 \cos 4t) + \frac{1}{10} (\cos t + 4 \sin t)$.

174. $y(t) = 18 - 17t - 18e^{-t} + 8t^2 - te^{-t}$.

Section 7.3

2. $y(x) = \frac{1}{24} \left[\frac{M}{EI\ell}\right]\{x^4 + 8\ell(x-\frac{1}{2}\ell)^3 u\ (x-\frac{1}{2}\ell) - 6\ell x^3 + 4\ell^3 x\}$, for $0 \le x \le \ell$.

4. $y(x) = \frac{1}{24} \left[\frac{M}{EI\ell}\right]\{x^4 = 2\ell x^3 + \ell^2 x^2\}$, for $0 \le x \le \ell$

6. $y(x) = \frac{1}{48} \left[\frac{M}{EI\ell}\right]\{3\ell^2 x - 4\ell x^3 + 8\ell(x-\frac{1}{2}\ell)^3 u\ (x-\frac{1}{2}\ell)\}$, for $0 \le x \le \ell$

8. $x_1(t) = \frac{1}{2} \cos \frac{1}{\sqrt{3}} t + \frac{3}{2} \cos t$

$x_2(t) = \frac{3}{2} \cos \frac{1}{\sqrt{3}} t - \frac{3}{2} \cos t$.

10. $x_1(t) = \frac{1}{2} e^{t}\cos t + \frac{1}{2} e^{-t} t$, $x_2(t) = \frac{1}{2} e^{t}\sin t - \frac{1}{2} e^{-t}\sin t$.

12. $x_1(t) = 1 - 2 \sin 3t$, $x_2(t) = 2 + \frac{2}{3} t + 2 \cos 3t$

14. $W(t) = e^{2t}$, (a) $y(t) = \frac{3}{4} (e^{2t} + \sin 2t - \cos 2t)$

(b) $y(t) = \frac{3}{4} (\sin 2t - \cos 2t - 3e^{-2t})$.

20. (i) $y(\infty) = 0$

(ii) $y(\infty)$ is not defined because zeros of denominator are purely
imaginary; in fact

$\mathcal{L}^{-1}\{Y(s)\} = \frac{1}{8}(\sin t - 2t \cos 2t)$.

22.   (i) $y(\infty)$ is not defined because denominator has positive zero ($s = 3/2$)

      (ii) $y(\infty) = -3/2$

24. $T(s) = \dfrac{sD}{K + sD}$

26. $T_\omega(s) = K_2/(K_1 K_2 + sJR)$, $T_i(s) = sJ/(K_1 K_2 + sJR)$.

30.   As, when a linear element is stable, a sinusoidal input produces a sinusoidal output, the result follows by setting $y_I = e^{i\omega t}$ and $y_O = Y(i\omega)e^{i\omega t}$ in the governing equation.

34. $e^{tA} = \mathscr{L}^{-1} \begin{bmatrix} \dfrac{s}{s^2 + 1} & \dfrac{-1}{s^2 + 1} & 0 \\[2mm] \dfrac{1}{s^2 + 1} & \dfrac{s}{s^2 + 1} & 0 \\[2mm] \dfrac{1}{(s - 1)(s^2 + 1)} & \dfrac{s}{(s - 1)(s^2 + 1)} & \dfrac{1}{s - 1} \end{bmatrix}$

but $\dfrac{1}{(s - 1)(s^2 + 1)} = \dfrac{1}{2}\left[\dfrac{1}{s - 1} - \dfrac{s}{s^2 + 1} - \dfrac{1}{s^2 + 1}\right]$ and

$\dfrac{s}{(s - 1)(s^2 + 1)} = \dfrac{1}{2}\left[\dfrac{1}{s - 1} - \dfrac{s}{s^2 + 1} + \dfrac{s}{s^2 + 1}\right]$,

so

$e^{tA} = \begin{bmatrix} \cos t & -\sin t & 0 \\ \sin t & \cos t & 0 \\ \frac{1}{2}(e^t - \cos = \sin t) & \frac{1}{2}(e^t - \cos t + \sin t) & e^t \end{bmatrix}$

$x_1(t) = \cos t - 2\sin t$

$x_2(t) = \sin t + 2\cos t$

$x_3(t) = \dfrac{5}{2} e^t - \dfrac{3}{2}\cos t + \dfrac{1}{2}\sin t.$

36. $x(t) = \sum\limits_{n=0}^{\infty} \dfrac{(-1)^n \, 2^n \, (t - 3n)^{n+1}}{(n + 1)!} u\,(t - 3n),$ for $t > 0.$

38. $x(t) = 2 \sum\limits_{n=0}^{\infty} \dfrac{(-1)^n (t - 2n)^{n+3}}{(n + 3)!} u\,(t - 2n),$ for $t > 0.$

40. $y(t) = t + e^t - 1 - \int_0^t (t - \tau)\tau\ y(\tau)\ d\tau.$

42. $y(t) = \frac{1}{6}\ t^3 - 2 \int_0^t (t - \tau)\ y(\tau)\ d\tau.$

44. $y(\tau) = 3t + \frac{1}{2}\ t^2 - \int_0^t \{3 - 2(t - \tau)\}\ y(\tau)\ d\tau.$

46. $y(t) = 1 + 2t - \sin t - \int_0^t \{1 - 3(t - \tau)\}y(\tau)\ d\tau.$

48. (i) $\int_0^t \frac{(t - \tau)^2}{2} \sin 3\tau\ d\tau$ (ii) $\int_0^t \frac{(t - \tau)^2}{2}\ e^{-3\tau}y(\tau)\ d\tau$

    (iii) $\int_0^t \frac{(t - \tau)^3}{6}\ (1 + \tau e^{-\tau})\ y(\tau)\ d\tau.$

50. $y(t) = 3t + \frac{3}{2}\ t^2 + \cos t - \int_0^t (t - \tau)^2 y(\tau)\ d\tau.$

52. $(1 + t)y'' + 2y' + (1 + t)y = 0$, with $y(0) = 1$, $y'(0) = 0$.

54. $\cos t\ y'' + ty' + (1 + t)y = 0$, with $y(0) = 0$, $y'(0) = 1$.

56. $y(t) = 1 + \sin t - \cos t.$

60. $[Y(s)]^2 - 9Y(s) + 9\left[\frac{3s - 1}{s^2}\right] = 0$, so $Y(s) = \frac{1}{s}$ or $Y(s) = 9 - \frac{3}{s}$.

    Thus $y(t) \equiv 3$ or $y(t) = 9\ \delta(t) - 3$. Only the first solution is finite for all time.

64. $y(t) = \frac{4}{3}\ (e^t - e^{-2t}).$

66. $y(t) = \frac{1}{3}\ e^t - \frac{1}{3}\ e^{-\frac{1}{2}t} \cos \frac{\sqrt{3}}{2}\ t + \frac{1}{\sqrt{3}}\ e^{-\frac{1}{2}t} \sin \frac{\sqrt{3}}{2}\ t.$

68. The transformed equation is

    $s(1 - s)Y' + (4 - s)Y = 0$

    with the solution

    $Y(s) = \frac{C(1 - s)^3}{s^4}.$

    The initial value theorem shows $C = 1$, and so

68. Cont'd...

$$Y(s) = \frac{(s-1)^3}{s^4} = \frac{1}{s} - \frac{3}{s^2} + \frac{3}{s^3} - \frac{1}{s^4}.$$

Hence $L_3(x) = \mathcal{L}^{-1}\{Y(s)\} = 1 - 3x + \frac{3}{2}x^2 - \frac{1}{6}x^3.$

70. The transformed equation is

$$Y' + (\tfrac{s}{2} + \tfrac{3}{s})Y = -1,$$

with the integrating factor $s^3 e^{s^2/4}$. Its solution is

$$Y(s) = \frac{8}{s^3} - \frac{2}{s} + \frac{A}{s^3} e^{-s^2/4},$$

where the first two terms on the right hand side are obtained by

evaluating $-\int s^3 e^{s^2/4} ds$ using integration by parts.

Expanding $s^{-3} e^{-s^2/4}$ and grouping terms in $1/s$ and $1/s^3$ gives

$$Y(s) = \frac{8}{s^3} - \frac{2}{s} + \frac{A}{s^3} - \frac{A}{4s} + \frac{A}{32}\left[s - \frac{s^3}{12} + ...\right].$$

The initial value theorem requires

$$-2 = \lim_{s\to\infty} (s\, Y(s)),$$

which implies $A = 0$. Hence $Y(s)$ and $H_2(x) = \mathcal{L}^{-1}\left\{\frac{8}{s^3} - \frac{2}{s}\right\} = $

$4x^2 - 2$ follow immediately.

72. The transformed equation is

$$(s^2 + 1)Y'' + 3sY' = 0.$$

One integration gives

$$\frac{dY}{ds} = \frac{A}{(s^2 + 1)^{3/2}},$$

and using the substitution $s = \sinh u$ a further integration gives

$$Y(s) = K + \frac{As}{(s^2 + 1)^{\frac{1}{2}}}.$$

Using the initial condition $J_1(0) = 0$ together with Theorem 7.15(i) shows

$K = -A$. The recurrence relations for $J_n(x)$ show $J_1'(0) = \frac{1}{2}$, and using

72. Cont'd...

this initial condition in Theorem 7.15(ii) shows K = 1, from which the result then follows.

74. Y(s) satisfies $2[-sY' - Y] + Y = -\frac{1}{2}\sqrt{\pi}\, s^{-3/2}\, e^{-\frac{1}{4s}}$,

or $Y' + \frac{1}{2s}\, Y = \frac{1}{4}\,\sqrt{\pi}\, s^{-5/2}\, e^{-\frac{1}{4s}}$.

This has the integrating factor $s^{\frac{1}{2}}$, so

$$\frac{d\,(s^{\frac{1}{2}}Y)}{ds} = \frac{1}{4}\,\sqrt{\pi}\, s^{-2}\, e^{-\frac{1}{4s}} \text{ and thus}$$

$$Y(s) = \sqrt{\pi}\, s^{-\frac{1}{2}}\, e^{-\frac{1}{4s}} + Cs^{-\frac{1}{2}} \quad (C = \text{const.})$$

For large s

$$\int_0^\infty e^{-st}\, \frac{\cos\sqrt{t}}{\sqrt{t}}\, dt \simeq \int_0^\infty e^{-st}\, t^{-\frac{1}{2}}\, dt = \sqrt{\pi}\, s^{-\frac{1}{2}}.$$

Comparing this with Y(s) for large s shows C = 0 so

$$\mathscr{L}\left\{\frac{\cos\sqrt{t}}{\sqrt{t}}\right\} = \sqrt{\pi}\, s^{-\frac{1}{2}}\, e^{-\frac{1}{4s}}, \text{ for } s > 0.$$

Section 7.4

2. $U(z) = z^{-1} + 2z^{-2} + 3z^{-3} + 4z^{-4}$.

4. $U(z) = 2 - \frac{3}{2}z^{-1} + \frac{4}{3}z^{-2} - \frac{5}{4}z^{-3} + \frac{6}{5}z^{-4} - \dots$.

6. $U(z) = a(Tz^{-1} + 2Tz^{-2} + 3Tz^{-3} + \dots + nTz^{-n} + \dots)$

and

$(z - 1)U(z) = aT(1 + z^{-1} + z^{-2} + \dots) = aT(1 - z^{-1})^{-1}$

so

$U(z) = \dfrac{aTz}{(z - 1)^2}$.

8. $U(z) = z^{-1} - z^{-3} + z^{-5} - z^{-7} + \ldots = z^{-1}(1 - z^{-2} + z^{-4} - z^{-6} + \ldots)$

$= z^{-1}(1 + z^{-2})^{-1} = \dfrac{z}{z^2 + 1}$.

10. $U(z) = (1 + z^{-1} + z^{-2} + z^{-3} + z^{-4} + z^{-5} + \ldots)$

$+ (z^{-1} - z^{-3} + z^{-5} - z^{-7} + \ldots)$

$= 1 + 2z^{-1} + z^{-2} + z^{-4} + 2z^{-5} + z^{-6} + z^{-8} + 2z^{-9} + \ldots$

$= (1 + z^{-2} + z^{-4} + \ldots) + 2z^{-1}(1 + z^{-2} + z^{-4} + \ldots)$

$= (1 - z^{-2})^{-1} + 2z^{-1}(1 - z^{-4})^{-1} = \dfrac{z^2(z + 1)}{(z - 1)(z^2 + 1)}$.

12. The result follows directly from $\cosh kt = \frac{1}{2}(e^{kt} + e^{-kt})$ and Theorem 7.19 by using $Z\{e^{kt}\}$ and $Z(e^{-kt})$.

14. Use $\cos^2 t = \frac{1}{2}(1 + \cos 2t)$ with $Z\{1\}$ and $Z\{\cos 2t\}$ to show

$U(z) = \dfrac{1}{2}\left[\dfrac{z}{z - 1} + \dfrac{z(z - \cos 2T)}{z^2 - 2z \cos 2T + 1}\right]$.

16. Use $\sinh t \cosh t = \frac{1}{2}\sinh 2t$ with $Z\{\sinh 2t\}$ to show

$U(z) = \dfrac{z \sinh 2T}{2(z^2 - 2z \cosh 2T + 1)}$.

18. Use $\sinh^2 t = \frac{1}{2}(\cosh 2t - 1)$ with $Z\{1\}$ and $Z\{\cosh 2t\}$ to show

$U(z) = \dfrac{1}{2}\left[\dfrac{z(z - \cosh 2T)}{z^2 - 2z \cosh 2T + 1} - \dfrac{z}{z - 1}\right]$.

20. $\dfrac{k}{s^2 + k^2} = \dfrac{k}{(s + ik)(s - ik)} = \dfrac{i}{2}\left[\dfrac{1}{s + ik} - \dfrac{1}{s - ik}\right]$, from which the result $Z\{\sin kt\}$ follows after using entry 7 of Table 7.2.

22. $\dfrac{k}{s^2 - k^2} = \dfrac{1}{2}\left[\dfrac{1}{s - k}\right] - \dfrac{1}{2}\left[\dfrac{1}{s + k}\right]$, so from entry 7 of Table 7.2

$U(z) = \dfrac{1}{2}\left[\dfrac{z}{z - e^{kt}}\right] - \dfrac{1}{2}\left[\dfrac{z}{z - e^{-kt}}\right] = \dfrac{z \sinh kT}{z^2 - 2z \cosh kt + 1}$.

24. $U(z) = \dfrac{1}{z - \frac{1}{2}} + \dfrac{2}{z - \frac{1}{3}}$, so $u_n = \left(\dfrac{1}{2}\right)^n + 2\left(\dfrac{1}{3}\right)^n$.

26. Identify with entry 14 of Table 7.2 by setting $e^{-2aT} = 4$ and

$2e^{-aT} \cos kT = 2\sqrt{2}$, so $e^{-aT} = 2$, $kT = \pi/4$. Thus

$\qquad U(z) = Z\{3e^{-at} \cos kt\}$ with $a = -(1/T) \ln 2$

and $k = \pi/4T$, so $u_n = 3 \cdot 2^n \cos \dfrac{n\pi}{4}$.

28. Identify with a linear combination of entries 13 and 14 of Table 7.2, with

$e^{-2aT} = 3$ and $2e^{-aT} \cos kT = \sqrt{6}$, so $e^{-aT} = \sqrt{3}$, $kT = \pi/4$. Thus

$\qquad U(z) \equiv AZ\{e^{-at} \sin kt\} + BZ\{e^{-at} \cos kt\}$

with $a = -(1/T) \ln \sqrt{3}$, $k = \pi/4T$. Equating the coefficients of powers of

z in the numerators of this identity shows $A = B = 1$, and hence that u_n

$= 3^{n/2} (\sin \dfrac{n\pi}{4} + \cos \dfrac{n\pi}{4})$.

34. $U(z) = z^{-2}V(z)$ with $V(z) = \dfrac{z^2}{(z - 2)(z - 3)} = \dfrac{3z}{z - 3} - \dfrac{2z}{z - 2}$.

Thus $v(nT) = 3^{n+1} - 2^{n+1}$ for $n = 0, 1, 2, \ldots$, but $Z\{u(t)\} =$

$Z\{v(t - 2T)\}$, so $u(nT) = 3^{n-1} - 2^{n-1}$ for $n = 2, 3, \ldots$ and $u(0) =$

$u(T) = 0$ because $v(-2T) = v(-T) = 0$.

36. $Z\{u(t + 3T)\} = z^3 U(z) - z^3 u(0) - z^2 u(T) - zu(2T)$ but $u(0) = 0$,

$u(T) = T^2 e^{-at}$, $u(2T) = 4T^2 e^{-2aT}$, so from entry 10 in Table 7.2,

$\qquad Z\{u(t + 3T)\} = \dfrac{T^2 e^4 (z + e^{-aT}) e^{-aT}}{(z - e^{-aT})^3} - T^2 e^{-aT} z^2 - 4T^2 e^{-2aT} z$.

38. $Z\{\sin 3t\} = \dfrac{z \sin 3T}{z^2 - 2z \cos 3T + 1}$, so $Z\{2^{t/T} \sin 3t\} =$

38. Cont'd...

$$\frac{(z/2) \sin 3T}{(z/2)^2 - 2(z/2) \cos 3T + 1} \; .$$

46. $u \; (0) = \lim_{z \to \infty} u(z) = \frac{2}{3} \; .$

48. Zeros of denominator all lie within unit circle $|z| = 1$.

$$\lim_{n \to \infty} u(nT) = \lim_{z \to 1} \left[\left[\frac{z-1}{z} \right] U(z) \right] = 0 \; .$$

50. Two zeros on the unit circle $|z| = 1$ at $z = \frac{1}{\sqrt{2}} \pm \frac{i}{\sqrt{2}}$ so limit not

defined.

52. $w(0) = 0$, $w(T) = 0$, $w(2T) = 1$, $w(3T) = 3$, $w(4T) = 5$, $w(5T) = 4$,

$w(6T) = 2$, $w(nT) = 0$, $n > 6$.

54. $U(z) = V(z) = Z\{u \; (t)\} - \frac{z}{z-1}$, so $W(z) = U(z)V(z) = \frac{z^2}{(z-1)^2}$,

$$w(kT) = \sum_{n=0}^{k} 1 = k + 1 \; .$$

56. $U(z) = \frac{z}{z-1}$, $V(z) = \frac{Tz}{(z-1)^2}$, so $W(z) = U(z)V(z) = \frac{Tz^2}{(z-1)^3} \; .$

$$w(kT) = \sum_{n=0}^{k} 1.(k - n)T = \frac{T}{2}k(1 - k).$$

Section 7.5

2. $U = \frac{az}{z-b}$, so $u_n = ab^n \; .$

4. $U = \frac{az^2 + (b - 2\alpha a)z}{(z - \alpha)^2} = \frac{az}{z - \alpha} + \frac{(b - \alpha a)z}{(z - \alpha)^2}$, so $u_n = a(1 - n)\alpha^n +$

$bn\alpha^{n-1} \; .$

6. $U = \dfrac{z^3 - 3z}{(z - 2)^2} + \dfrac{2z}{(z - 1)(z - 2)^2} = \dfrac{2z}{z - 1} - \dfrac{z}{z - 2} + \dfrac{z}{(z - 2)^2}$,

so $u_n = 2 - 2^n + n2^{n-1}$.

8. $U = \dfrac{z^2 - 3z}{(z - 1)(z - 2)} + \dfrac{z}{(z - 1)(z - 2)(z - 4)} = \dfrac{7}{3}\left[\dfrac{z}{z - 1}\right] -$

$\dfrac{3}{2}\left[\dfrac{z}{z - 2}\right] + \dfrac{1}{6}\left[\dfrac{z}{z - 4}\right]$, so $u_n = \dfrac{7}{3} - \left[\dfrac{3}{2}\right]2^n + \left[\dfrac{1}{6}\right]4^n$.

10. $U = \dfrac{z}{z^2 - \sqrt{2}z + 1}$. Identification with entry 11 of Table 7.2 with

$\cos kT = 1/\sqrt{2}$, so $kT = \pi/4$, shows

$u_n = \sqrt{2} \sin \dfrac{n\pi}{4}$.

12. $zU = 3V + \dfrac{z}{z - 1}$ and $zV = 3U - \dfrac{z}{z - 1}$, giving

$U = \dfrac{z^2 - 3z}{(z - 1)(z^2 - 9)} = \dfrac{1}{4}\left[\dfrac{z}{z - 1}\right] - \dfrac{1}{4}\left[\dfrac{z}{z - 3}\right]$,

so $u_n = \dfrac{1}{4} - \dfrac{1}{4}(-3)^n$, $v_n = \dfrac{1}{3}(u_{n+1} - 1)$.

14. $G(s) = \dfrac{2}{3}\left[\dfrac{1}{s + 1}\right] + \dfrac{1}{3}\left[\dfrac{1}{s + 4}\right]$, so $G(z) = \dfrac{2}{3}\left[\dfrac{z}{z - e^{-T}}\right] + \dfrac{1}{3}\left[\dfrac{z}{z - e^{-4T}}\right]$.

Now $Y_1(z) = Z\{e^{-t} \cos t\} = \dfrac{z(z - e^{-T} \cos T)}{z^2 - 2ze^{-T} \cos T + e^{-2T}}$,

so from (22) we have $Y_0(z) = G(z) Y_1(z)$.

16. $G_1(s) = \dfrac{1}{(s + 2)^2 + 1}$, so $G_1(z) = \dfrac{ze^{-T} \sin 2T}{z^2 - 2ze^{-T} \cos 2T + e^{-2T}}$.

Hence $G_s(z) = G_1(z) G_2(z) = \dfrac{z^2 e^{-T} \sin 2T}{(z - e^{-3T})(z^2 - 2ze^{-T} \cos 2T + e^{-2T})}$.

Chapter 8.   Series Solution of Ordinary Differential Equations.

Section 8.1

2. $a_n = 1/(2n + 1)$; $r = 1$; $-1 < x < 1$.

4. $a_n = (-1)^n/\{(2n + 1)(2n + 1)!\}$; $r = \infty$; $-\infty < x < \infty$.

6. $a_n = n^n$; $r = 0$; series only converges at the origin .

8. $a_n = (4n)^n/\{n^n + 3\}$; $r = 4$; $-4 < x < 4$.

10. Both f and g have radius of convergence 1 and interval of convergence

    $0 < x < 2$.   The sum $f(x) + g(x) = \displaystyle\sum_{n=1}^{\infty} \left[\frac{1}{n^{1/2}} + \frac{1}{n^3}\right](x - 1)^n$ has as its

    interval of convergence $0 < x < 2$.

12. $\sin(x^2)\ \sinh(2x) = 2x^3 + \frac{4}{3}x^5 - \frac{1}{15}x^7 + \ldots$.

14. $\mathrm{sech}\ x = (\cosh x)^{-1} = \left[1 + \left[\frac{x^2}{2!} + \frac{x^4}{4!} + \frac{x^6}{6!} + \ldots\right]\right]^{-1}$

    $= 1 - \left[\frac{x^2}{2!} + \frac{x^4}{4!} + \frac{x^6}{6!} + \ldots\right] + \left[\frac{x^2}{2!} + \frac{x^4}{4!} + \frac{x^6}{6!} + \ldots\right]^2 + \ldots$

    $= 1 - \frac{1}{2}x^2 + \frac{5}{24}x^4$ (up to terms in x^4).

16. $\sin x\ \exp\{1 - \cos x\} = x + \frac{1}{3}x^3 + \ldots$.

18. The result follows directly by mathematical induction using the differen-
    tiability property of a power series within its interval of convergence.

20. $d(\sec x)/dx = \sec x \tan x = x + \frac{5}{6}x^3 + \frac{61}{120}x^5 + \frac{277}{1008}x^7 + \ldots$,

    for $\frac{-\pi}{2} < x < \frac{\pi}{2}$.

22. $\frac{1}{3} d[(\arcsin x)^3]/dx = \dfrac{(\arcsin x)^2}{\sqrt{1 - x^2}} = x^2 + \frac{3!}{4!}\ 3\ (1 + \frac{1}{3^2})x^4$

    $+ \frac{3!}{6!}\ 3.5^2\ (1 + \frac{1}{3^2} + \frac{1}{5^2})x^6 + \ldots$; $r = 1$, $-1 < x < 1$.

24. $f'(x) = \dfrac{1}{1 + x^2} = 1 - x^2 + x^4 - x^6 + \ldots$ (by the Binomial theorem).

    So $f(x) = \arctan x = \displaystyle\int_0^x \frac{1}{1 + t^2}\ dt = x - \frac{x^3}{3} + \frac{x^5}{5} - \frac{x^7}{7} + \ldots$.

26. $\text{erf } x = \dfrac{2}{\sqrt{\pi}} \displaystyle\int_0^x \{1 - t^2 + \dfrac{t^4}{2!} - \dfrac{t^6}{3!} + ...\} dt = \dfrac{2x}{\sqrt{\pi}} \{1 - \dfrac{x^2}{1!3} + \dfrac{x^4}{2!5} - \dfrac{x^6}{3!7}$

$+...\}$, $-\infty < x < \infty$.

28. $1 - 2x^2 + \dfrac{10}{3} x^4$.

30. $\dfrac{1}{x} + \dfrac{x}{6} + \dfrac{7}{360} x^3 + \dfrac{31}{15,120} x^5 + \dfrac{127}{604,800} x^7$.

32. (i) $\displaystyle\sum_{m=1}^{\infty} (m + 2)(m + 4)x^{m+1}$ (ii) $\displaystyle\sum_{m=1}^{\infty} (m + 2)(m + 5)x^{m+1}$.

34. $(1 + x^2) \displaystyle\sum_{n=3}^{\infty} n(n + 1)x^{n-2} = \displaystyle\sum_{m=0}^{\infty} (m + 2)(m + 3)x^m + \displaystyle\sum_{n=3}^{\infty} n(n + 1)x^n$

$= 6 + 12x + 20x^2 + \displaystyle\sum_{n=3}^{\infty} [(n + 2)(n + 3) + n(n + 1)]x^n$

$= 6 + 12x + 20x^2$

$+ \displaystyle\sum_{r=2}^{\infty} [(r + 3)(r + 4) + (r + 1)(r + 2)]x^{r+1}$.

36. $F(x) = \displaystyle\sum_{n=2}^{\infty} n(n - 1)a_n x^n + \displaystyle\sum_{n=1}^{\infty} 3na_n x^{n-1}$

$= \displaystyle\sum_{n=2}^{\infty} n(n - 1)a_n x^n + \displaystyle\sum_{m=2}^{\infty} 3(m - 1)a_{m-1} x^{m-2}$

$= \displaystyle\sum_{n=2}^{\infty} n(n - 1)a_n x^n + \displaystyle\sum_{r=0}^{\infty} 3(r + 1)a_{r+1} x^r$

$= 3a_1 + 6a_2 x + \displaystyle\sum_{n=2}^{\infty} [n(n - 1)a_n + 3(n + 1)a_{n+1}]x^n$.

38. The series for x cot x follows from the series for x coth x by replacing x by ix because $\sinh(ix) = i\sin x$ and $\cosh(ix) = \cos x$. The series for tan x then follows by multiplying the identity by x and combining terms in the series for xcot x and 2xcot (2x).

42. $\int_0^1 \dfrac{te^{xt}}{e^t - 1} \, dx = 1 = \sum_{n=0}^{\infty} \left[\int_0^1 B_n(x)dx \right] \dfrac{t^n}{n!}$, and as this must be true for

all t it follows that

$$\int_0^1 B_n(x)dx = 0 \quad \text{for } n = 1, 2, \dots .$$

$B_1'(x) = B_0(x) = 1$, so $B_1(x) = x + c$. However $\int_0^1 B_1(x)dx = 0$,

so $\int_0^1 (x + c)dx = 0$, whence $c = -\dfrac{1}{2}$ and hence $B_1(x) = x - \dfrac{1}{2}$.

Similar reasoning gives

$$B_2(x) \text{ to } B_4(x) .$$

Section 8.2

2. $y(x) = 3 + 4x + \dfrac{5}{2} x^2 + x^3 + \dfrac{7}{24} x^4 + \dots .$

4. $y(x) = 3 - 2x + \dfrac{3}{2} x^2 - \dfrac{1}{3} x^3 + \dfrac{1}{8} x^4 - \dfrac{1}{60} x^5 + \dots$

6. $y(x) = 2 - 3x^2 + \dfrac{9}{4} x^4 - \dfrac{27}{24} x^6 + \dots$

8. $y(x) = (x - \dfrac{\pi}{2}) + \left[\dfrac{12 - 8\pi + 8\pi^2 + \pi^4}{24} \right] (x - \dfrac{\pi}{2})^3 + \dots$

12. $y(x) = 2 + 2x - \dfrac{2}{3} x^3 - \dfrac{1}{3} x^4 + \dots .$

14. $y(x) = 1-(x - 1) - \dfrac{1}{4}(x - 1)^2 + \dfrac{1}{6} (x - 1)^3 - \dfrac{5}{96} (x - 1)^4 \dots$

16. $y(x) = 2 - x^2 + \dfrac{1}{3} x^3 - \dfrac{1}{10} x^4 + \dfrac{1}{30} x^5 - \dots$

18. $y(x) = x - \dfrac{1}{3} x^3 + \dfrac{1}{12} x^4 + \dfrac{1}{15} x^5 + \dots .$

20. $y(x) = 1 + (x - 1) + \dfrac{1}{12} (x - 1)^4 - \dfrac{2}{15} (x - 1)^5 +$

$\dfrac{59}{360} (x - 1)^6 + \dots$

Section 8.3

2. $2a_2 + 3a_0 = 0$, $(n + 1)(n + 2)a_{n+2} - (n - 3)a_n = 0$, $n = 1, 2, \dots$,

so $\quad a_2 = -\dfrac{3}{2}a_0$, $a_4 = \dfrac{3}{4!}a_0$, $a_6 = \dfrac{1.3}{6!}a_0$, $a_8 = \dfrac{1.3.5}{8!} a_0, \dots$,

2. Cont'd...

and $a_3 = -\frac{1}{3}a_1$, $a_{2n+1} = 0$, $n = 2, 3, \ldots$.

$$y(x) = a_0\left[1 - \frac{3}{2}x^2 + \frac{3}{4!}x^4 + \frac{1.3}{6!}x^6 + \frac{1.3.5}{8!}x^8 + \ldots\right]$$
$$+ a_1\left[x - \frac{1}{3}x^3\right] .$$

4. $2a_2 + a_0 = 0$, $6a_3 + a_1 = 0$, $(n + 1)(n + 2)a_{n+2} + (n^2 - n + 1)a_n$
$= 0$, $n = 2, 3, \ldots$.

$$y(x) = a_0\left[1 - \frac{1}{2}x^2 + \frac{1}{8}x^4 - \frac{13}{240}x^6 + \ldots\right]$$

$$+ a_1\left[x - \frac{1}{6}x^3 + \frac{7}{120}x^5 - \frac{7}{240}x^7 + \ldots\right]$$

6. $4a_2 + a_0 = 0$, $2(n + 1)(n + 2)a_{n+2} - (4n - 1)a_n = 0$, $n = 1, 2, \ldots$.

$$a_{2n} = \frac{-(4.2 - 1)(4.4 - 1)(4.6 - 1) \ldots [4(2n - 2) - 1]}{2^n(2n)!}a_0 ,$$

$$a_{2n+1} = \frac{(4.1 - 1)(4.3 - 1)(4.5 - 1) \ldots [4(2n - 1) - 1]}{2^n(2n + 1)!}a_1,$$

$n = 1, 2, \ldots$.

$$y(x) = a_0\left[1 - \frac{1}{2.2!}x^2 - \frac{1.7}{2^2.4!}x^4 - \frac{1.7.15}{2^3.6!}x^6 - \ldots\right]$$
$$+ a_1\left[x + \frac{1}{2.3!}x^3 + \frac{1.5}{2^2.5!}x^5 + \frac{1.5.9}{2^3.7!}x^7 + \ldots\right] .$$

8. Write $x^2 = 1 - 2(x + 1) + (x + 1)^2$ so the differential equation becomes

$$[1 - 2(x + 1) + (x + 1)^2]y'' + y = 0 .$$

$$2a_2 + a_0 = 0 , \ 6a_3 + 4a_2 + a_1 = 0 ,$$

$$(n + 1)(n + 2)a_{n+2} - 2n(n + 1)a_{n+1} + (n^2 - n + 1)a_n = 0 ,$$

$$n = 2, 3 \ldots .$$

$$y(x) = a_0\left[1 - \frac{1}{2}(x + 1)^2 + \frac{11}{24}(x + 1)^4 + \frac{13}{30}(x + 1)^5 + \ldots\right]$$
$$+ a_1\left[(x + 1) - \frac{1}{6}(x + 1)^3 - \frac{1}{6}(x + 1)^4 - \frac{1}{12}(x + 1)^5 + \ldots\right] .$$

10. $a_2 + 10a_0 = 0$, $a_3 + 3a_1 = 0$

$$(n + 1)(n + 2)a_{n+2} - (n^2 - n - 20)a_n = 0, \ n = 2, 3, \ldots$$

71

10. Cont'd...

$$y(x) = a_0\left[1 - 10x^2 + \frac{35}{3}x^4\right] + a_1\left[x - 3x^3 + \frac{6}{5}x^5 + \frac{2}{7}x^7 + \dots\right].$$

The first linearly independent solution reduces to a quartic.

12. $a_2 = 0$, $6a_3 + a_0 = 1$, $(n + 1)(n + 2)a_{n+2} + a_{n-1} = 0$, $n = 2, 3, \dots$

$$y(x) = a_0\left[1 - \frac{1}{2.3}x^3 + \frac{1}{2.3.5.6}x^6 - \dots\right]$$

$$+ a_1\left[x - \frac{1}{3.4}x^4 + \dots\right]$$

$$+ \left[\frac{1}{2.3}x^3 - \frac{1}{2.3.5.6}x^6 + \dots\right].$$

14. $2a_1 + a_0 = 0$, $2a_{n+1} + a_n = 0$, $n = 1, 2, \dots$

$$a_n = \frac{(-1)^n}{2^n}a_0 ,$$

$$y(x) = a_0\left[1 - \frac{1}{2}x + \frac{1}{4}x^2 - \frac{1}{8}x^3 + \dots\right].$$

Compare this with the analytic solution $y = c/(2 + x)$.

16. Write $x = (x + 2) - 2$ so the differential equation becomes

$$[(x + 2) - 2]y' + (x + 2)^2 y = 0 .$$

$a_1 = a_2 = 0$, $na_n - 2(n + 1)a_{n+1} + a_{n-2} = 0$, $n = 2, 3, \dots$

$$y(x) = a_0\left[1 + \frac{1}{6}(x + 2)^3 + \frac{1}{16}(x + 2)^4 + \frac{1}{40}(x + 2)^5 + \frac{7}{288}(x + 2)^6 + \dots\right].$$

Section 8.4

2. Routine calculations using (12).

4. Follows from (12) by writing the terms in the reverse order (decreasing powers of x) and removing the factor $(2n)!/[2^n(n!)^2]$.

6. Routine calculations.

8. $f(x) = \frac{28}{35}P_0(x) + \frac{13}{5}P_1(x) + \frac{16}{7}P_2(x) + \frac{2}{5}P_3(x) + \frac{32}{35}P_4(x)$, $-1 \le x \le 1$.

10. Establish the first result as indicated in the problem. The Taylor series expansion of $G(x, r)$ about the origin is

$$G(x, r) = G(x, 0) - \left.\frac{\partial G}{\partial r}\right|_{r=0} x + \left.\frac{\partial^2 G}{\partial r^2}\right|_{r=0} \frac{x^2}{2!} + \left.\frac{\partial^3 G}{\partial r^3}\right|_{r=0} \frac{x^3}{3!} + \cdots \ ,$$

and after using the first result this becomes

$$G(x, r) = \sum_{n=0}^{\infty} P_n(x) x^n \ .$$

12. Proceed as indicated in the problem.

14. $\displaystyle \phi = \frac{Q}{r_2} - \frac{Q}{r_1} = \frac{Q}{r\left[1 - 2\left(\frac{d}{r}\right)\cos\theta + \left(\frac{d}{r}\right)^2\right]^{1/2}}$

$$- \frac{Q}{r\left[1 - 2\left(\frac{d}{r}\right)\cos(\pi - \theta) + \left(\frac{d}{r}\right)^2\right]^{1/2}} \ .$$

Expand the first term using the result of Problem 13 and the second using the same result with $\cos\theta$ replaced by $\cos(\pi - \theta) = -\cos\theta$. The result follows by combining the two series. The potential of the dipole follows by taking the limit.

16. Routine calculations.

18. The result $P_n(1) = 1$ follows by induction from (26) by setting $x = 1$ and using the fact that $P_0(1) = P_1(1) = 1$. The result $P_n(-1) = (-1)^n$ follows in similar fashion by setting $x = -1$ in (26), using induction, and the fact that $P_0(-1) = 1$ and $P_1(-1) = -1$.

20. Integrating by parts n times gives

$$\int_{-1}^{1} Q_m(x) P_n(x) dx = \frac{1}{2^n n!} \int_{-1}^{1} Q_m(x) \frac{d^n}{dx^n}[(x^2 - 1)^n] dx$$

$$= \frac{(-1)^n}{2^n n!} \int_{-1}^{1} (x^2 - 1)^n \frac{d^n}{dx^n} Q_m(x) \ dx \ .$$

20. Cont'd...

If $Q_m(x)$ is a polynomial of degree $m < n$ then $\dfrac{d^n Q_m(x)}{dx} = 0$, and
the first result follows. The second result follows by identifying $Q_m(x)$
with $P_m(x)$ when $m < n$ and by interchaning $P_m(x)$ and $P_n(x)$ when
$m > n$.

Section 8.5

6. $3^6 \Gamma(6\tfrac{2}{3})/\Gamma(\tfrac{2}{3})$.

8. $3^6 \Gamma(7\tfrac{1}{3})/\Gamma(1\tfrac{1}{3})$.

10. $\dfrac{\Gamma(n + \tfrac{1}{2})}{4^n \Gamma(\tfrac{1}{2})} x^n$.

12. $\dfrac{1}{3^{2n}} \dfrac{\Gamma(n + \tfrac{1}{3})}{\Gamma(\tfrac{1}{3})} x^n$.

14. $\dfrac{(-1)^n 2^\alpha \Gamma(\alpha + 1) x^{2n+\alpha}}{2^{2n} \Gamma(n + 1) \Gamma(\alpha + n + 1)}$.

16. (i) $\Gamma(x + 1) = \lim\limits_{n \to \infty} \left[\dfrac{nx}{x + n} \right] \cdot \dfrac{1.2 \ldots (n - 1)}{x(n + 1) \ldots (x + n - 1)} n^x = x\Gamma(x)$.

(ii) Result follows directly from (i) when n is a positive integer.

(iii) $\Gamma(1) = \lim\limits_{n \to \infty} \dfrac{n!}{n!} = 1$.

18. $\Gamma(p)\Gamma(1 - p) = \left[\int_0^\infty u^{p-1} e^{-u} du \right] \left[\int_0^\infty v^{-p} e^{-v} dv \right]$

$= 4 \int_0^\infty x^{2p-1} e^{-x^2} dx \int_0^\infty y^{1-2p} e^{-y^2} dy$

$= 4 \int_0^{\pi/2} \int_0^\infty r e^{-r^2} (\cot \theta)^{2p-1} d\theta$

$= 2 \int_0^{\pi/2} (\cot \theta)^{2p-1} d\theta$.

Setting $z = \cot^2 \theta$ this becomes

$\Gamma(p)\Gamma(1 - p) = \int_0^\infty \dfrac{z^{p-1}}{1 + z} dz$,

from which the required result follows after using the stated result for the
integral.

74

20. The result follows by repeated application of the first result of Problem 19 coupled with the fact that

$$B(p, 1) = \int_0^1 x^{p-1}dx = \frac{1}{p} .$$

The second result follows from the first by setting $p = m$, multiplying numerator and denominator by $(m - 1)!$, and using the definition of the gamma function.

22. In the second integral set $t = 1/x$ to obtain

$$B(m, n) = \int_0^1 \frac{t^{m-1}}{(1 + t)^{m+n}} dt ,$$

and then replace t by x.

Section 8.6

2. Regular singular points at $x = \pm 2n\pi$, $n = 0, 1, 2, \ldots$.

4. Regular singular points at $x = 0$ and $x = -2$ and a regular singular point at infinity.

6. Irregular singular point at $x = 0$ and a regular singular point at infinity.

8. $t(t + 3)y'' + 2(t + 3)y' + (t + 4)y = 0$.

10. $y(x) = c_1 x^{1/2} \sum_{n=0}^{\infty} \frac{(-1)^n x^n}{2^n 2!} + c_2 \left\{ 1 + \sum_{n=1}^{\infty} \frac{(-1)^n x^n}{1.3.5.7 \ldots (2n - 1)} \right\}$.

12. $y(x) = c_1 \left\{ 1 + \sum_{n=1}^{\infty} \frac{(-1)^n x^{2n}}{2^n n! \ 3.7.11 \ldots (4n - 1)} \right\}$

$\quad + c_2 x^{1/2} \left\{ 1 + \sum_{n=1}^{\infty} \frac{(-1)^n x^{2n}}{2^n n! \ 5.9.13 \ldots (4n + 1)} \right\}$.

14. $y(x) = c_1 \sum_{n=0}^{\infty} \frac{(-1)^n x^n}{(2n)!} + c_2 x^{1/2} \sum_{n=0}^{\infty} \frac{(-1)^n x^n}{(2n + 1)!}$.

16. Set $z = 1/x$, solve the resulting equation for z, and then write $z = 1/x$.

$$y(x) = c_1 x^{-1/2} \sum_{n=0}^{\infty} \frac{(-1)^n}{2^n} \left[\frac{1}{x}\right]^n + c_2 \left[1 + \sum_{n=1}^{\infty} \frac{(-1)^n}{1.3.5.7 \ldots (2n-1)} \left[\frac{1}{x}\right]^n\right].$$

18. $$y(x) = c_1 \left[1 + \sum_{n=1}^{\infty} \frac{(-1)^n (2n-5)(2n-7) \ldots 3.1.1.3 \ x^{2n}}{(4n-3)(4n-5) \ldots 9.5}\right]$$

$$+ c_2 x^{3/2} \left[1 + \sum_{n=1}^{\infty} \frac{(-1)^{n+1}(2n-7)(2n-9) \ldots 3.1.1.3.5 \ x^{2n}}{2^{3n} n!}\right].$$

20. $$y_1(x) = 1 - \frac{1}{2x^2} + \frac{1}{8x^4} + \ldots \ ,$$

$$y_2(x) = \frac{1}{x}\left[1 - \frac{1}{3x^2} + \frac{1}{15x^4} + \ldots\right].$$

22. $$y_1(x) = \sum_{n=0}^{\infty} \frac{x^n}{n!(n+1)!} \ , \quad \text{and} \quad y_2(x) = \frac{1}{x} + y_1 \ln|x| + \frac{1}{2} - \frac{1}{2}x + \ldots$$

26. $$y(x) = x^{1/2}\left[c_1 x^{\sqrt{3/4}} + c_2 x^{-\sqrt{3/4}}\right].$$

Section 8.7

10. Multiply (20) by $x^{\upsilon-1}$ to arrive at the result

$$x^{\upsilon} J_{\upsilon}' = x^{\upsilon} J_{\upsilon-1} - \upsilon x^{\upsilon-1} J_{\upsilon} \ , \quad \text{or}$$

$$x^{\upsilon} J_{\upsilon-1} = x^{\upsilon} J_{\upsilon}' + \upsilon x^{\upsilon-1} J_{\upsilon} = \frac{d}{dx}(x^{\upsilon} J_{\upsilon}).$$

The other result follows in similar fashion after multiplying (19) by $x^{-(\upsilon+1)}$.

20. $y(x) = c_1 J_2(3x) + c_2 Y_3(3x)$.

22. $y(x) = c_1 J_{1/3}(x\sqrt{2}) + c_2 J_{-1/3}(x\sqrt{2})$.

24. Set $u = dy/dx$, then

$$\frac{dy}{dx} = c_1 J_{1/2}(2x) + c_2 J_{-1/2}(2x), \text{ so}$$

$$y(x) = c_1 \int J_{1/2}(2x)dx + c_2 \int J_{-1/2}(2x)dx + c.$$

26. $x^3 J_3(x) + c$.

28. $-x^{-1} J_1(x) + c$.

30. $J_0(x) - x^{-1} J_1(x) + c$.

32. $x^2 J_2(x) + J_1(x) + c$.

34. Integrate (16) and use the result $J_0(0) = 1$.

36.
$$\cos(x \sin \theta) = 1 - \frac{x^2 \sin^2\theta}{2!} + \frac{x^4 \sin^4\theta}{4!} - \cdots ,$$

so

$$\frac{1}{\pi} \int_0^\pi \cos(x \sin \theta)d\theta = \frac{1}{\pi}\left[\int_0^\pi d\theta - \frac{x^2}{2}\int_0^\pi \sin^2\theta d\theta + \frac{x^4}{2^4}\int_0^\pi \sin^4\theta d\theta - \cdots\right]$$

$$= \frac{1}{\pi}\left[\pi - \frac{x^2}{2}\cdot\frac{\pi}{2} + \frac{x^4}{24}\cdot\frac{3\pi}{8} - \cdots\right]$$

from which the required result then follows.

46.
$$\frac{1}{2}(t - \frac{1}{t})\exp[\frac{1}{2}x(t - \frac{1}{t})] = \sum_{n=-\infty}^{\infty} t^n J_n'(x) ,$$

which may be written

$$\frac{1}{2}\sum_{n=-\infty}^{\infty} [t^{n+1} J_n(x) - t^{n-1} J_n(x)] = \sum_{n=-\infty}^{\infty} t^n J_n'(x) .$$

Equating the coefficient of t^n on either side of this identity gives the required result.

Section 8.8

8. $ci(x) \sim \cos x\left[\dfrac{1}{x^2} - \dfrac{3!}{x^4} + \dfrac{5!}{x^6} - \cdots\right] - \sin x\left[\dfrac{1}{x} - \dfrac{2!}{x^3} + \dfrac{4!}{x^5} - \cdots\right]$.

10. $c(x) \sim \cos x^2\left[\dfrac{1}{4x^3} - \dfrac{15}{16x^7} + \cdots\right] - \sin x^2\left[\dfrac{1}{2x} - \dfrac{3}{8x^5} + \cdots\right]$.

Use $C(x) = \displaystyle\int_0^\infty \cos^2 t\, dt - \int_x^\infty \cos^2 t\, dt = \frac{1}{2}\sqrt{\frac{\pi}{2}} - c(x)$.

18. $\int_0^\infty \dfrac{e^{-xt}\ln(1+t)}{1+t}\,dt \sim \dfrac{1}{x^2} - \dfrac{3}{x^3} + \dfrac{11}{x^4} - \cdots$.

22. The integral is in the correct form with $a(t) = -[(t-2)^2 + 1]$ and
    $f(t) = (1+t^2)^{-1}$. $a(t)$ has a maximum at $t = 2$ at which $a(2) = -1$
    and $a''(2) = -2$. Thus from the Laplace formula

    $$I(x) \sim \dfrac{1}{5}\sqrt{\dfrac{\pi}{x}}\,e^{-x} .$$

24. $[t^2 - 8t + 19]^{-n} = \exp[-n\,\ln[(t-4)^2 + 3]]$, so $a(t) =$
    $-\ln[(t-4)^2 + 3]$ and $f(t) = 1$.

    The function $a(t)$ has a maximum at $t = 4$, with $a(4) = -\ln 3$ and $a''(4) =$
    $-2/3$. Thus from the Laplace formula $I_n \sim \sqrt{\dfrac{3\pi}{n}}\dfrac{1}{3^n}$. Alternatively, use

    the result of Problem 23 with $p(x) = [(t-4)^2 + 3]^{-1}$.

28. The equation is in the standard form

    $$y_+(x) \sim x^{-1/4}\exp\left[\tfrac{2}{3}x^{3/2}\right] , \quad y_-(x) \sim x^{-1/4}\exp\left[-\tfrac{2}{3}x^{3/2}\right] .$$

30. $y = ue^{x/2}/x^{3/4} , \quad u'' - \dfrac{1}{4}\left[1 + \dfrac{7}{x} - \dfrac{3}{4x^2}\right]u = 0$.

    $$y_+(x) \sim x\exp\left[x + \dfrac{13}{4x}\right] , \quad y_-(x) \sim x^{-5/2}\exp\left[-\dfrac{13}{4x}\right] .$$

32. The equation is in standard form, but here $f(x) = -x$ with $x > 0$ so
    complex quantities arise (see Problem 26).

    $$y_+(x) \sim x^{-1/4}\cos\left[\tfrac{2}{3}x^{3/2}\right] , \quad y_-(x) \sim x^{-1/4}\sin\left[\tfrac{2}{3}x^{3/2}\right] .$$

34. Divide by x to put in the standard form, giving

    $$y'' - \left[1 + \dfrac{2}{x}\right]y = 0, \quad \text{so } f(x) = \dfrac{2}{x} + 1 .$$

    $$y_+(x) \sim \left[x - 2 + \dfrac{5}{8x} + 0(x^{-2})\right]\exp\left[x + \dfrac{1}{x}\right] ,$$

    $$y_-(x) \sim \left[\dfrac{1}{x} - \dfrac{2}{x^2} + \dfrac{5}{8x^3} + 0(x^{-4})\right]\exp\left[-(x + \dfrac{1}{x})\right] .$$

Section 8.9

2.

n	x_n	k_{1n}	K_{1n}	k_{2n}	K_{2n}
0	1	0.2	0.227015	0.211351	0.214298
1	1.1	0.221395	0.201277	0.231459	0.189603
2	1.2	0.240329	0.177751	0.249217	0.167041
3	1.3	0.257013	0.156234	0.264824	0.146400
4	1.4	0.271636	0.136526	0.278463	0.127484

n	x_n	k_{3n}	K_{3n}	k_{4n}	K_{4n}
0	1	0.210715	0.213436	0.221344	0.201241
1	1.1	0.230876	0.188910	0.240286	0.177726
2	1.2	0.248681	0.166478	0.256977	0.156217
3	1.3	0.264333	0.145939	0.271606	0.136513
4.	1.4	0.278011	0.127103	0.284347	0.118432

n	x_n	k_n	K_n	y_n	z_n
0	1	0.210912	0.213954	0.0	2.0
1	1.1	0.231059	0.189338	0.210912	2.213954
2	1.2	0.248850	0.166834	0.441971	2.403292
3	1.3	0.264489	0.146237	0.690821	2.570126
4	1.4	0.278155	0.127355	0.955310	2.716364

4.

n	x_n	k_{1n}	K_{1n}	k_{2n}	K_{2n}
0	0	0.0	0.350000	0.017500	0.367500
1	0.1	0.036686	0.383007	0.055836	0.398289
2	0.2	0.076417	0.410432	0.096030	0.421883
3	0.3	0.118453	0.428222	0.139864	0.433019
4.	0.4	0.161522	0.429765	0.183010	0.423729

n	x_n	k_{3n}	K_{3n}	k_{4n}	K_{4n}
0	0	0.018375	0.366610	0.036661	0.382918
1	0.1	0.056600	0.396980	0.076384	0.410335
2	0.2	0.097511	0.419920	0.118409	0.428118
3	0.3	0.140104	0.430106	0.161463	0.429665
4	0.4	0.182708	0.419569	0.203479	0.405727

4. Cont'd...

n	x_n	k_n	K_n	y_n	z_n
0	0	0.018068	0.366856	0.5	0.0
1	0.1	0.056324	0.397313	0.518068	0.366856
2	0.2	0.097288	0.420359	0.574329	0.764169
3	0.3	0.139975	0.430690	0.671680	1.184528
4	0.4	0.182740	0.420348	0.811655	1.615218

6.

n	x_n	k_{1n}	K_{1n}	k_{2n}	K_{2n}
0	2	0.0	0.625000	0.039063	0.597293
1	2.125	0.074318	0.560857	0.109371	0.522799
2	2.25	0.139375	0.477270	0.169204	0.432645
3	2.375	0.193429	0.389290	0.217760	0.353484
4	2.5	0.237889	0.329794	0.258501	0.320435

n	x_n	k_{3n}	K_{3n}	k_{4n}	K_{4n}
0	2	0.037331	0.593251	0.074156	0.561165
1	2.125	0.106993	0.519323	0.139233	0.477641
2	2.25	0.166415	0.431203	0.193275	0.389641
3	2.375	0.215522	0.353899	0.237667	0.329999
4	2.5	0.257916	0.321600	0.278089	0.329983

n	x_n	k_n	K_n	y_n	z_n
0	2	0.037824	0.594542	−1.0	0.0
1	2.125	0.107713	0.520457	−0.962176	0.594542
2	2.25	0.167315	0.432434	−0.854463	1.115000
3	2.375	0.216277	0.355676	−0.687148	1.547434
4	2.5	0.258135	0.323975	−0.470871	1.903110

Chapter 9. Fourier Series, Sturm–Liouville Problems and Orthogonal
        Functions.

Section 9.1

2. 4π 4. 6π 6. 8π

8.   0   10. π

12.

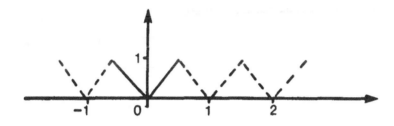

14.

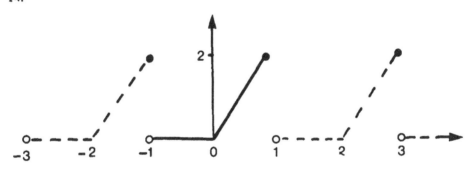

16.

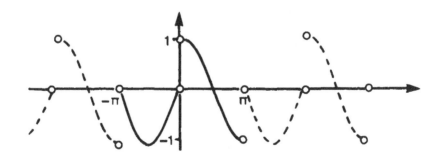

18. Both extensions are possible because f(0)=0.

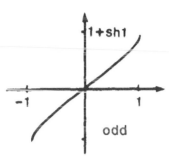

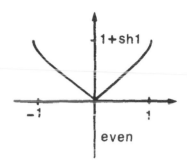

20. Both extensions are possible because f(0)=0.

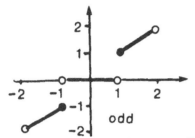

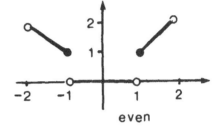

22. (a) piecewise continuous (b) piecewise smooth (c) piecewise smooth.

24. $0 < a < 1$ it follows that the limit function $f(x) = \lim_{n\to\infty} f_n(x) \equiv 0$ for $0 \leq x \leq a$. Thus $|f_n(x) - f(x)| = x^{2n}$, and this may be made arbitrarily small for all x in the interval $0 \leq x \leq a$ by taking n sufficiently large, thereby establishing the uniform convergence on this interval. The uniform convergence fails in the event that $a = 1$ because the limit function becomes discontinuous, as in Example 9.2.

26. The Maclaurin series for sinh x is

$$\sinh x = \sum_{n=1}^{\infty} \frac{x^{2n+1}}{(2n+1)!} \ .$$

Thus

$$|s_n(x) - \sinh x| = |\sum_{r=n+1}^{\infty} \frac{x^{2r+1}}{(2r+1)!}| \leq \sum_{r=n+1}^{\infty} \frac{a^{2r+1}}{(2r+1)!}$$

for $-a \leq x \leq a$. The last summation represents the sum of the tail of

82

26. Cont'd...

an infinite series which is convergent by the ratio test, and so it can be made arbitrarily small by taking n sufficiently large for $-a \le x \le a$. The uniform convergence is thus established, and the result remains true for any value of a.

Section 9.2

2.

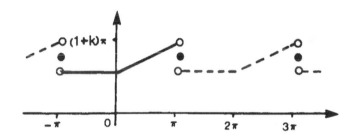

4.

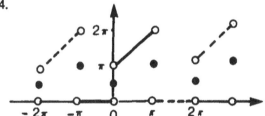

Series converges to $\frac{\pi}{2}$ at x=0, $\sum_{n=1}^{\infty} \frac{1}{(2n-1)^2} = \frac{\pi^2}{8}$.

Series converges to $\frac{3\pi}{2}$ at $x=\frac{\pi}{2}$, $\sum_{n=1}^{\infty} \frac{(-1)^{n+1}}{(2n-1)} = \pi/4$.

6.

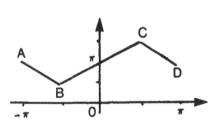
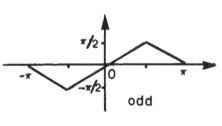

Function is continuous so at $x=\pi/2$ the series converges to $3\pi/2$.

$$\sum_{n=1}^{\infty} \frac{1}{(2n-1)^2} = \frac{\pi^2}{8}.$$

8.

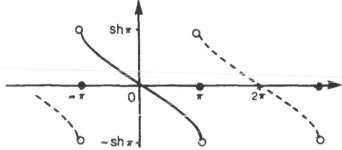

10.

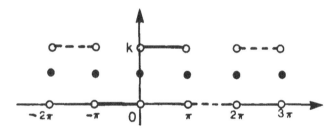

12.

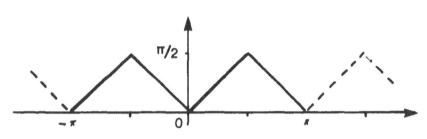

14.

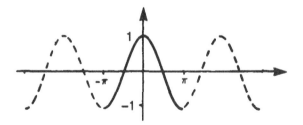

Set $k\pi = \alpha$ and $x = \pi$. Then at $x = \pi$ the series converges to $\cos \alpha$.

Divide by $\sin \alpha$ and substitute for k in terms of α to obtain the result.

16.

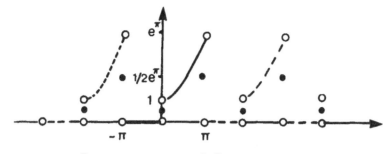

$$f(x) \sim \frac{e^{\pi}-1}{2\pi} + \frac{1}{\pi} \sum_{n=1}^{\infty} \left[\frac{(-1)^n e^{\pi}-1}{1+n^2} \right] (\cos nx - n \sin nx).$$

At $x = 0$ the Fourier series converges to $\frac{1}{2}$, and at $x = \pi$ to $\frac{1}{2} e^{\pi}$. Thus

$$\frac{1}{2} = \frac{e^{\pi}-1}{2\pi} + \frac{1}{\pi} \sum_{n=1}^{\infty} \left[\frac{(-1)^n e^{\pi}-1}{1+n^2} \right] \text{ and}$$

$$\frac{1}{2} e^{\pi} = \frac{e^{\pi}-1}{2\pi} + \frac{1}{\pi} \sum_{n=1}^{\infty} \left[\frac{e^{\pi}-(-1)^n}{1+n^2} \right].$$

Multiplying by $\pi e^{-\pi/2}$ and grouping even and odd terms gives

$$\frac{\pi}{2} e^{-\pi/2} = \sinh \frac{\pi}{2} - 2 \cosh \frac{\pi}{2} A + \sinh \frac{\pi}{2} B$$

and

$$\frac{\pi}{2} e^{\pi/2} = \sinh \frac{\pi}{2} + 2 \cosh \frac{\pi}{2} A + \sinh \frac{\pi}{2} B,$$

where

$$A = \sum_{n=0}^{\infty} \frac{1}{1+(2n+1)^2} = \sum_{n=0}^{\infty} \frac{1}{4n^2+4n+2} \text{ and } B = \sum_{n=1}^{\infty} \frac{1}{1+4n^2}.$$

Eliminating first A, then B between the equations gives the required sums.

18.

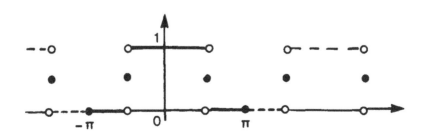

20.

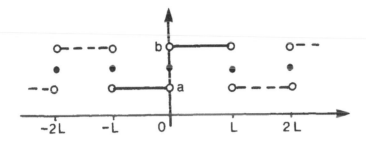

22.

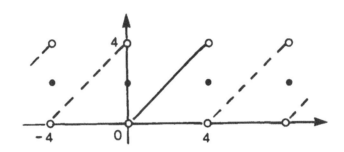

24.

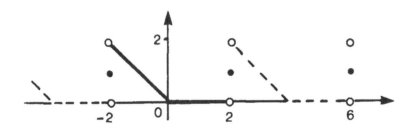

26.

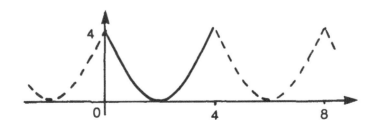

28.

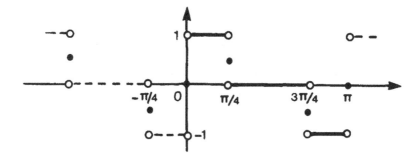

30.

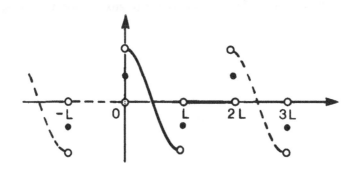

32.

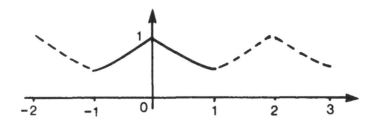

$$f(x) \sim \frac{1}{e} - \frac{2}{e} \sum_{n=1}^{\infty} \left[\frac{e - (-1)^n}{1 + n^2\pi^2}\right] \cos n\pi x + 4 \sum_{n=1}^{\infty} \left[\frac{1}{2n-1}\right] \sin (2n-1)\pi x$$

$$- \frac{2\pi}{e} \sum_{n=1}^{\infty} \left[\frac{e - (-1)^n}{1 + n^2\pi^2}\right] \sin n\pi x .$$

The required result follows by setting $x = 0$, when the Fourier series converges to the value zero.

34.

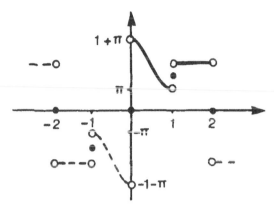

$$f(x) \sim \pi + \frac{1}{2} \cos \pi x + \frac{4}{\pi} \sum_{n=1}^{\infty} \left[\frac{n}{4n^2 - 1} \right] \sin 2nx , \quad \text{for } 0 < x \leq 2.$$

36.

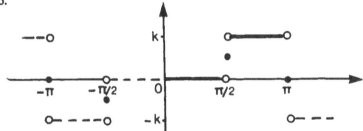

$$f(x) \sim \frac{2k}{\pi} \sum_{n=1}^{\infty} \left[\frac{\cos \frac{n\pi}{2} + (-1)^{n+1}}{n} \right] \sin nx , \quad \text{for } 0 \leq x < \pi.$$

38.

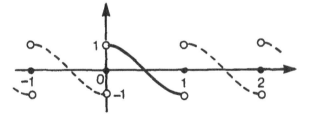

In the interval $(k, k+1)$ the series represents a periodic extension of $f(x)$. At $x = 1/4$ the function is continuous, so at this point the series converges to $\cos \pi/4 = 1/\sqrt{2}$. This leads to the result

$$\sum_{n=0}^{\infty} \frac{(-1)^n (2n+1)}{4(2n+1)^2 - 1} = \frac{\pi}{8\sqrt{2}} .$$

40.

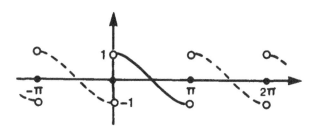

42.

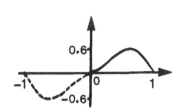

odd extension

$$f(x) \sim \frac{1}{2} \sin \pi x - \frac{16}{\pi^2} \sum_{n=1}^{\infty} \frac{n}{(4n^2-1)^2} \sin 2n\pi x,$$

for $0 \leq x \leq 1$.

even extension

$$f(x) \sim \frac{1}{\pi}\left[1 - \frac{1}{2} \cos \pi x + 2 \sum_{n=2}^{\infty} \frac{(-1)^{n-1}}{(n^2-1)} \cos n\pi x\right],$$

for $0 \leq x \leq 1$.

44.

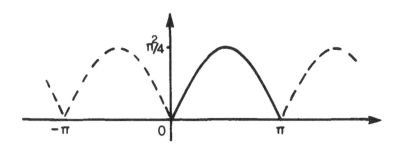

46. When $k = N\pi/L$ the series remains unchanged apart from replacing c_N in the summation by $i/2$ and c_{-N} by $-i/2$.

89

48.

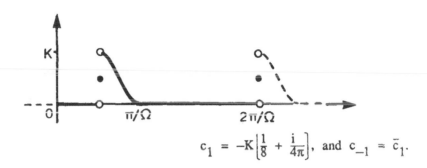

$$c_1 = -K\left[\frac{1}{8} + \frac{i}{4\pi}\right], \text{ and } c_{-1} = \bar{c}_1.$$

50.

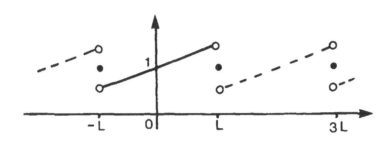

$$c_n = (-1)^n \frac{\sinh \alpha L \ (\alpha L + in\pi)}{\alpha^2 L^2 + n^2 \pi^2} \ , \ c_{-n} = \bar{c}_n \text{ and}$$

$$f(x) \sim \lim_{m\to\infty} \sum_{n=-m}^{m} c_n \, e^{in\pi x/L} \ .$$

Set $\alpha=L=1$ and combine terms to obtain the representation of $f(x) = e^{-x}$
for $-1 < x < 1$.

52. $F(x) = 1-3 \ f(x) \ ; \ F(x) \sim -\frac{1}{2} - \frac{6}{\pi} \sum_{n=0}^{\infty} \frac{\sin(2n+1)(\pi x/6)}{(2n+1)} \ .$

54. $F(x) = 1 + f(x) - g(x) \ ; \ F(x) \sim 1 + \frac{2}{\pi} \sum_{n=0}^{\infty} \frac{\sin(2n+1)\pi x}{(2n+1)}$

$$- \frac{4}{\pi} \sum_{n=1}^{\infty} \frac{(-1)^n \cos n\pi x}{n^2} \ .$$

56. $F(x) = \cos 2x -1+4f(x) \ ; \ F(x) \sim 1+\cos 2x+\frac{8}{\pi}\sum_{n=0}^{\infty} \frac{\sin(2n+1)x}{(2n+1)} \ ,$ because

cos 2x is its own Fourier series representation.

58.

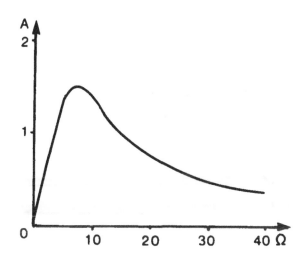

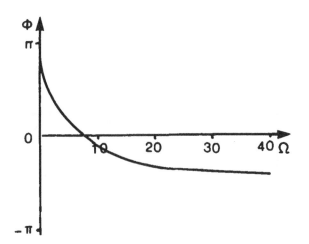

T(0) = −0.082

T(i) = −0.078 + 0.27i

T(2i) = 0.071 + 0.547i

T(3i) = 0.33 + 0.753i

T(4i) = 0.671 + 0.841i

60.

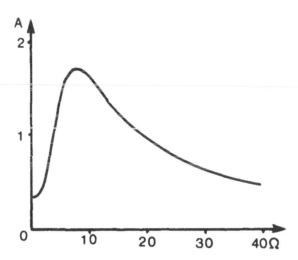

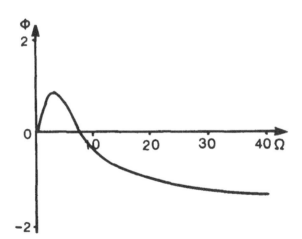

T(0) = 0.62

T(i) = 0.563 + 0.053i

T(2i) = 0.55 + 0.253i

T(3i) = 0.727 + 0.455i

T(4i) = 1.016 + 0.505i

Section 9.3

2. $\displaystyle\sum_{n=1}^{\infty} \frac{1}{(2n-1)^4} = \frac{\pi^4}{96}$.

4. $f(x) \sim \dfrac{1}{\pi} + \dfrac{1}{2}\sin x - \dfrac{2}{\pi}\displaystyle\sum_{n=1}^{\infty} \dfrac{\cos 2nx}{(2n-1)(2n+1)}$

from which the result follows by application of the Parseval relation.

6. The Fourier series for $f(x)$ only contains cosines so all its coefficients b_n vanish. The Fourier series for $g(x)$ only contains sines, so all its coefficients A_n vanish. Consequently, one factor vanishes in each of the products in the series

$$\frac{1}{2}a_0 A_0 + \sum_{n=1}^{\infty} (a_n A_n + b_n B_n),$$

so that

$$\int_{-L}^{L} f(x)\ g(x)\ dx = 0.$$

Section 9.4

2. The first result follows from Theorem 9.10 by setting $L=\pi$ and $c = -\pi$.

$$f(x) \sim 2 \sum_{n=1}^{\infty} \frac{(-1)^{n+1}}{n}\sin nx.$$

Applying Theorem 9.10 gives

$$\int_{-\pi}^{x} t\ dt = 2\sum_{n=1}^{\infty} \frac{(-1)^n}{n} \int_{-\pi}^{x} \sin nt\ dt.$$

So

$$x^2 - \pi^2 = 4\sum_{n=1}^{\infty} \left[\frac{(-1)^n(\cos nx - (-1)^n)}{n^2}\right],$$

from which the required result follows after using the known sum

$$\sum_{n=1}^{\infty} \frac{1}{n^2} = \frac{\pi^2}{6} .$$

4. $f(x) \sim \pi - 2 \sum_{n=1}^{\infty} \frac{\sin nx}{n}$. An application of Theorem 9.10 gives

$$\int_0^x t \, dt = \int_0^x \pi \, dt - 2 \sum_{n=1}^{\infty} \frac{1}{n} \int_0^x \sin nt \, dt,$$

from which the result follows after using the known series

$$\sum_{n=1}^{\infty} \frac{1}{n^2} = \frac{\pi^2}{6} .$$

6. $f(x)$ is continuous with $f(-\pi) = f(\pi)$ so Theorem 9.11 applies and shows that $g(x) = |\cos x|$, for $-\pi \le x \le \pi$.

8. $f(x)$ is continuous with $f(-2) = f(2)$ so Theorem 9.11 applies and yields for the Fourier series representation of $g(x)$

$$g(x) \sim \frac{4}{\pi} \sum_{n=0}^{\infty} \frac{\sin (2n+1)(\pi x/2)}{(2n+1)} .$$

Section 9.7

2.

n	0	1	2	3
a_n	3.574	0.729	0.357	−0.191
b_n	–	−1.606	−0.234	0.098

4.

n	0	1	2	3
a_n	2.76	1.266	0.462	−0.438
b_n	–	0.312	−0.249	0.033

6.

n	0	1	2	3
a_n	3.399	0.727	0.374	−0.284
b_n	–	−1.395	−0.437	0.019

94

8.

n	0	1	2	3
a_n	2.769	0.677	0.343	0.007
b_n	–	–0.023	–1.237	–0.017

Section 9.8

2. $y(x) = A \cos 3\sqrt{\lambda}\, x + B \sin 3\sqrt{\lambda}\, x$.

($y(0) = 0$) $A = 0$ ($y(L) = 0$) $\sin 3\sqrt{\lambda}\, L = 0$ thus

$3\sqrt{\lambda}\, L = n\pi$, so the eigenvalues are $\lambda_n = \dfrac{n^2\pi^2}{9L^2}$ and the eigenfunctions are

$y_n(x) = \sin \dfrac{n\pi x}{L}$, $n = 1, 2, \ldots$.

4. $y(x) = A \cos\sqrt{\lambda}\, x + B \sin\sqrt{\lambda}\, x$, so $y'(x) = -A\sqrt{\lambda}\, \sin\sqrt{\lambda}\, x +$

$B\sqrt{\lambda}\, \cos\sqrt{\lambda}\, x$.

($y'(-\pi) = 0$) $A \sin\pi\sqrt{\lambda} + B \cos\pi\sqrt{\lambda} = 0$

($y(\pi) = 0$) $A \cos\pi\sqrt{\lambda} + B \sin\pi\sqrt{\lambda} = 0$.

Non–trivial solution if $\begin{vmatrix} \sin\pi\sqrt{\lambda} & \cos\pi\sqrt{\lambda} \\ \cos\pi\sqrt{\lambda} & \sin\pi\sqrt{\lambda} \end{vmatrix} = 0$

which is equivalent to $\cos 2\pi\sqrt{\lambda} = 0$.   Thus $2\pi\sqrt{\lambda} = (2n-1)\dfrac{\pi}{2}$, and so

the eigenvalues are $\lambda_n = (\dfrac{2n-1}{4})^2$, for $n = 1, 2, \ldots$.   Now $A/B =$

$-\cot \pi\sqrt{\lambda_n}$, so setting $B = -1$ (it is arbitrary) gives for the eigenfunctions

$y_n(x) = \cos(\dfrac{2n-1}{4})\pi \cos (\dfrac{2n-1}{4})x - \sin (\dfrac{2n-1}{4})x,\ n = 1, 2, \ldots$.

6. $y(x) = A \cos\sqrt{\lambda}\, x + B \sin\sqrt{\lambda}\, x$, so $y'(x) = -A\sqrt{\lambda}\, \sin\sqrt{\lambda}\, x$

$+ B\sqrt{\lambda}\, \cos \sqrt{\lambda}\, x$.

($y(0) = y(2\pi)$) $A = A \cos 2\pi\sqrt{\lambda} + B \cos 2\pi\sqrt{\lambda}$

($y'(0) = y'(2\pi)$) $B = -A \sin 2\pi\sqrt{\lambda} + B \cos 2\pi\sqrt{\lambda}$.

Non–trivial solution if $\begin{vmatrix} (1-\cos 2\pi\sqrt{\lambda}) & -\sin 2\pi\sqrt{\lambda} \\ \sin 2\pi\sqrt{\lambda} & (1-\cos 2\pi\sqrt{\lambda}) \end{vmatrix} = 0$

6. Cont'd...

which is equivalent to $\cos 2\pi\sqrt{\lambda} = 1$. Thus $2\pi\sqrt{\lambda} = 2n\pi$, and so the eigenvalues are $\lambda_n = n^2$, for $n = 0, 1, 2, \ldots$. For these eigenvalues the arbitrary constants A and B are no longer related, so the eigenfunctions become 1, $\cos x$, $\cos 2x$, $\ldots$, corresponding to the arbitrary constant A, and $\sin x$, $\sin 2x$, $\ldots$, corresponding to the arbitrary constant B. Thus the representation of $f(x)$ becomes the ordinary Fourier series representation based on these trigonometric functions.

Section 9.9

2. $\lambda_n = \mu_{1,n}^2$, $y(x) = J_1(\mu_{1,n} x)$, $f(x) = x$ so

$$x = \sum_{n=1}^{\infty} c_n J_1(\mu_{1,n} x) \text{ , with}$$

$$c_n = \frac{1}{\|y_n\|^2} \int_0^1 f(x) \, x \, J_1(\mu_{1,n} x) \, dx = \frac{1}{\|y_n\|^2} \int_0^1 x^2 J_1(\mu_{1,n} x) \, dx \ .$$

$$\|y_n\|^2 = \frac{1}{2} [J_2(\mu_{1,n})]^2 \ , \quad c_n = \frac{1}{\|y_n\|^2} \frac{1}{(\mu_{1,n})^3} \int_0^{\mu_{1,n}} x^2 J_1(x) \, dx$$

$$= \frac{2}{[J_2(\mu_{1,n})]^2} \cdot \frac{1}{(\mu_{1,n})^3} \cdot (\mu_{1,n})^2 J_2(\mu_{1,n}) = \frac{2}{\mu_{1,n} J_2(\mu_{1,n})} \ .$$

4. $\lambda_n = \mu_{0,n}^2$, $y_n(x) = J_0(\mu_{0,n} x)$, $f(x) = \begin{cases} 0, & 0 \le x < a \\ 1, & a < x < b \\ 0, & b < x \le 1 \end{cases}$ so

$$f(x) \sim \sum_{n=1}^{\infty} c_n J_0(\mu_{0,n} x) \text{ , with}$$

4. Cont'd...

$$c_n = \frac{1}{\|y_n\|^2} \int_0^1 f(x) \; x \; J_0 \; (\mu_{0,n} \; x)dx = \frac{1}{\|y_n\|^2} \int_a^b x \; J_0 \; (\mu_{0,n} \; x)dx \;\; .$$

$$\|y_n\|^2 = \int_0^1 [J_0(\mu_{0,n} \; x)]^2 \; xdx = \tfrac{1}{2} \; [J_1 \; (\mu_{0,n})]^2 \;\; .$$

$$c_n = \frac{2}{[J_1(\mu_{0,n})]^2} \int_0^n x \; J_0(\mu_{0,n}x)dx = \frac{2}{[J_1(\mu_{0,n})]^2} \; \frac{1}{(\mu_{0,n})^2} \int_{\mu_{0,n}a}^{\mu_{0,n}b} x \; J_0(x) \; dx$$

$$= \frac{2}{[J_1(\mu_{0,n})]^2} \; \frac{1}{(\mu_{0,n})^2} \; [\mu_{0,n}b \; J_1 \; (\mu_{0,n}b) - \mu_{0,n}a \; J_1 \; (\mu_{0,n}a)]$$

$$= \frac{2}{\mu_{0,n}\{J_1(\mu_{0,n})\}^2} \; [b \; J_1 \; (\mu_{0,n} \; b) - a \; J_1 \; (\mu_{0,n} \; a)] \;\; .$$

6. As the interval $a \leq x \leq b$ $(b > a > 0)$ does not contain the origin the general solution must contain both $J_0(kx)$ and $Y_0(kx)$, so that

$$y(x) = A \; J_0(kx) + BY_0(kx) \;\; .$$

$(y(a) = 0)$ $\quad 0 = A \; J_0(ka) + BY_0(ka)$

$(y(b) = 0)$ $\quad 0 = A \; J_0(kb) + BY_0(kb) \;\; .$

Non–trivial solutions if

$$\begin{vmatrix} J_0(ka) & Y_0(ka) \\ J_0(kb) & Y_0(kb) \end{vmatrix} = 0 \;\; .$$

Thus k must be a zero of the transcendental equation

$$J_0(ka) \; Y_0(kb) - J_0(kb) \; Y_0(ka) = 0 \;\; .$$

With these values of k, say k_1, k_2, ... ,

$$\frac{A}{B} = - \frac{Y_0(k_n a)}{J_0(k_n a)} \;\; ,$$

6. Cont'd...

so as B is arbitrary, the eigenfunctions may be taken to be

$$y_n(x) = J_0(k_n a) \, Y_0(k_n x) - Y_0(k_n a) \, J_0(k_n x) \, .$$

As this is a Sturm–Liouville problem the eigenfunctions $y_n(x)$ are orthogonal over the interval $a \le x \le b$ with weight function x. Thus

$$f(x) = \sum_{n=1}^{\infty} c_n \, y_n(x) \, , \quad \text{with} \quad c_n = \frac{1}{\|y_n\|^2} \int_a^b f(x) \, y_n(x) \, x \, dx \, .$$

Section 9.10

6. Using (18), express $T_n''(x)$ and $T_n'(x)$ in terms of $T_n(x)$, and then show all three expressions satisfy the Chebyshev differential equation.

12. Using (28), express $L_n''(x)$ and $L_n'(x)$ in terms of $L_n(x)$, and then show all three expressions satisfy the Laguerre differential equation.

18. Integrate by parts and use the result $\int_{-\infty}^{\infty} e^{-x^2} \, dx = \sqrt{\pi}$.

20. Using (35), express $H_n''(x)$ and $H_n'(x)$ in terms of $H_n(x)$, and then show all three expressions satisfy the Hermite differential equation.

98

Printed and bound by CPI Group (UK) Ltd, Croydon, CR0 4YY

22/10/2024

01777600-0012